Deviations in Contemporary Theatrical Anthropology

This book refers to the artistic deviation from dominant goals in a social system or from means considered legitimate in that system.

This book explores a "New Humanism" in the performing arts, unique in the sense of human's ability to co-create and communicate beyond spatial and temporal boundaries, wars, and pandemics, through artistic deviations carried out by machines and through the Extended Reality. Through the lens of anthropology and aesthetics, this study selects useful case studies to demonstrate this phenomenon of performative sympoiesis, in which the experimentation of AI-driven creativity and the new human-robot interaction (HRI) lead to philosophical inquiries about the nature of creativity, intelligence, and the definition of art itself. These shifts in paradigms invite us to reconsider established concepts and explore new perspectives on the relationship between technology, art, and the human experience.

This book will be of great interest to students and scholars in theatre and performance studies, anthropology, and digital humanities.

Ester Fuoco is a Senior Researcher in Performing Arts at IULM University. In 2019, she obtained a PhD in Digital Humanities from the University of Genoa through a joint program with Paris 7 University. In 2021, she received a research grant at the Institute of Biorobotics at the Sant'Anna School of Advanced Studies. From 2022 to 2023, she was a postdoctoral researcher on a project titled "Onlife Theatre: Creative and Reception Convergences between Live, Digital, and Transmediality". Her main research areas focus on the use of new technologies in performance from an anthropological and aesthetic perspective.

Routledge Advances in Theatre & Performance Studies

This series is our home for cutting-edge, upper-level scholarly studies and edited collections. Considering theatre and performance alongside topics such as religion, politics, gender, race, ecology, and the avant-garde, titles are characterized by dynamic interventions into established subjects and innovative studies on emerging topics.

Logomimesis
A Treatise on the Performing Body
Esa Kirkkopelto, translated by Kate Sotejeff-Wilson

Performing Climates
Eddie Paterson and Lara Stevens

Mess and Contemporary Performance
Complexity, Containment, and Collapse
Harriet Curtis

Performance
The Ethics and the Politics of Conservation and Care, Volume II
Edited by Hanna B. Hölling, Julia Pelta Feldman, Emilie Magnin

Samuel Beckett's Italian Modernisms
Tradition, Texts, Performance
Edited by Michela Bariselli, Davide Crosara, Antonio Gambacorta, and Mario Martino

Deviations in Contemporary Theatrical Anthropology
New Myths and Performative Rituals between XR, Robots, and AI
Ester Fuoco

For more information about this series, please visit: www.routledge.com/Routledge-Advances-in-Theatre--Performance-Studies/book-series/RATPS

Deviations in Contemporary Theatrical Anthropology

New Myths and Performative Rituals between XR, Robots and AI

Ester Fuoco

LONDON AND NEW YORK

First published 2025
by Routledge
4 Park Square, Milton Park, Abingdon, Oxon OX14 4RN

and by Routledge
605 Third Avenue, New York, NY 10158

Routledge is an imprint of the Taylor & Francis Group, an informa business

British Library Cataloguing-in-Publication Data
A catalogue record for this book is available from the British Library

ISBN: 9781032676906 (hbk)
ISBN: 9781032676913 (pbk)
ISBN: 9781032676937 (ebk)

DOI: 10.4324/9781032676937

Typeset in Times New Roman
by codeMantra

To my mother who has always given me the courage to dare and persevere

Contents

Introduction

Nowadays, there is talk of intelligent or autonomous machines, infallible and fast, capable of replacing humans in many of their activities, even artistic ones. However, theater, dramaturgical creativity linked to fantasy, imagination, and even intentionality seemingly remain human capacities, although not equally developed in all. In this sense, the title of this book refers to "deviation" whose sociological connotation pertains to "the deviation from dominant goals in a social system or from the means considered legitimate in that system".[1]

What makes this transition to a New Humanism in the Performing Arts unique is undoubtedly man's ability to co-create and communicate beyond spatial and temporal boundaries, wars, and pandemics, through artistic deviations realized by the machines he has invented and developed. Art has never stopped and now—technological art—has become "viral and vital", with all the implications that these adjectives entail. In this book, we will attempt to support an ever-evolving cartography related to various technological performative experiments. Through the lens of anthropology and aesthetics, we have selected case studies useful to demonstrate this phenomenon of performative sympoiesis,[2] in which AI-driven creativity experimentation and the new interaction between humans and machines lead to philosophical inquiries into the nature of consciousness, intelligence, and the very definition of art. These paradigm shifts invite us to reconsider established concepts and explore new perspectives on the relationship between technology, art, and human experience. Anthropology, akin to philosophy, does not focus on objects but examines the way of observing these objects and subjects this to analysis. Therefore, this volume seeks to overlay this analytical filter with the aesthetic one, looking 'through' not by quickly moving from one discipline to another, but by attempting to pour some assumptions of one theory into another, creating a dialogue on the fundamental question of this research: *whether it is possible to think about theater and define what it truly is.*

Almost half a century ago, Clifford Geertz[3] noted a great tumultuous change taking place in the social sciences, a path branching in various directions, aiming to outline a new design of the cultural map and a modification of

DOI: 10.4324/9781032676937-1

the underlying principles. This writing constitutes the beginning of a process in this direction: a shift in the way we conceptualize the thinking process regarding theater. Anthropology, which has always identified itself with the hermeneutic model and centers its study on living subjects captured in their ability to produce meaning through their actions, serves as a methodological model for analyzing these uncertain, fluid performative forms, for which an appropriate lexicon is sometimes lacking.

In this period defined as the "fourth revolution", new technologies[4]—specifically information and communication technologies (ICT)—have undoubtedly influenced and altered our sense of self, our ways of relating, and the processes through which we shape and interact with the world. Man's self-understanding has been disrupted, as have the concepts of identity, experience, history, time, and space; our entire experience seems to be migrating toward another intangible world made of connections. As philosopher Luciano Floridi explains, billions of people interact directly or indirectly. Through Facebook, a million and a half people use YouTube every day, and almost the entire world population communicates via mobile phones. The phenomenon of the Internet of Things (IoT) demonstrates that in the last ten years, there have been more "objects" connected to each other than human communications, with *machine-to-machine*[5] interactions prevailing.

In this dystopian scenario, we investigate the intrinsic anthropological value of ancient art such as theater, an art whose birth coincides with ritual and has its roots in the remotest antiquity. The following questions—and perhaps the apprehension—arise: whether the artist will be replaced by a machine or an avatar, and whether physical presence, a specific element of the performing arts, will be supplanted by the virtual one. In this discourse, we will try to understand how the field of performing arts has been (and is) a protagonist of paradoxical transformations, becoming increasingly an object of mediated and remote enjoyment. Perhaps we still lack adequate categories or definitions, conceptually and linguistically, for the rapid evolution of the hybridization phenomena we witness daily, but it is possible to try to identify some positive aspects of this technologization, alongside the more critical ones.

Performing arts, therefore, find themselves at the crossroads of this ontological enigma. They have always been about the representation of life, but in this era, they become a form of life in themselves, a hybrid of atoms and bits. This phenomenon extends the concept of mimesis, as the very act of representation intertwines with the form of digital mediation. Are we, in our essence, simply flesh algorithms, or is there an irreducible quality of "humanity" that resists digitization? This investigation into performative sympoiesis—a co-creation intertwined between human and machine—highlights the dialectic between freedom and determinism, between the narrative self and the connected self, challenging the very foundations on which we build our understanding of consciousness, intelligence, and art.

Returning to the source, how can we think about theater? Certainly, with one eye on the past and one on the future, but in this case, we dare to present it in its current state, as it is. Indeed, a historical or philological analysis has already been extensively conducted, despite the fact that archives and records are never a completely exhaustive or objective method as sometimes perceived;[6] they indeed provide a partial portrait of all the possibilities that theater can represent, perhaps sometimes with a bit of presumption. Concerning the intent of this book, directed towards those who open their minds to unexplored landscapes and data, we would like to say that what theater should be carries a distinctly anthropological character, to arrive at what reality is; hence, that world and that magma of interdisciplinary performance studies that somehow diverge from what traditional theater is.[7] Considering theater in its current form represents a significant challenge, requiring a direct confrontation with the utopias, experiments, failures, and biases that characterize it. Reflecting on theater today implies not only evaluating its traditions and established practices but also embracing the innovation and constant evolution that define its contemporary panorama.

Theater has always been fertile ground for experimentation. Theatrical utopias, often idealistic, have pushed the boundaries of what is possible on stage, imagining new ways to engage audiences and represent reality. These experiments, although not always successful, have played a crucial role in the development of theatrical language, paving the way for new forms of expression and communication. Theatrical experiments, ranging from the integration of new technologies to experimenting with unconventional narrative forms, represent bold attempts to reinvent the medium.[8]

In the context of live performing arts, as suggested by anthropological study, the phenomenon under examination is approached not only from the front or behind. This perspective allows for a more holistic and multidimensional understanding of the performance, recognizing the active role of all elements involved—actors, spectators, and technology—in creating a dynamic and interactive theatrical experience. Such an approach emphasizes the importance of examining theater not only as an observed event but also as a field of live and reciprocal interaction, where the boundary between stage and audience blurs, giving rise to a new dimension of dialogue and participation.

Their interest particularly focuses on the actor, the spectator, and technology, seen as active bodies within the theater under analysis. This approach allows for the examination of performances as moments of encounter comparable to artistic events, seeking to assess what a technological show can offer and how much the spectator is able to perceive. Flusser's philosophy,[9] centered on the transformative impact of technology on society and culture, finds direct application in the context of theater and intermedial performance. These artistic fields act as microcosms to concretely explore and experiment with Flusser's theories on the interaction between humanity and technology.

In contemporary theater, especially in intermedial performance that combines various media such as sound, video, and virtual reality, a tangible manifestation of Flusser's "telematic society" can be observed. These experimental theatrical forms challenge traditional codes of storytelling and representation, much like Flusser's "technical images" alter visual communication. In intermedial performance, artists use technology not only as a scenic tool but as an essential component of creative expression, thus redefining the relationship between the actor, the spectator, and the stage space.

Technology in intermedial theater can be used to create immersive environments that transform the perception of space and time, breaking physical barriers and allowing spectators to enter a digitally constructed world. This aligns with Flusser's concept of a culture where physical distance is rendered irrelevant by digital connectivity. This artistic practice raises critical questions about the authenticity of human experience in an era where reality can be simulated or completely constructed through digital means.

Moreover, theater and intermedial performances directly confront Flusser's dialectic of freedom and determinism. In theater incorporating digital elements, live action can be manipulated, distorting reality through technological mediation. This pushes artists and audiences to question the nature of art and its ability to reflect human reality or create an alternative reality.

Intermedial performance[10] thus becomes a testing ground for Flusser's "performative sympoiesis", where co-creation between humans and technology occurs in real time. This hybrid of live performance and advanced technology reflects the tensions and possibilities explored in Flusser's philosophy, serving as a scenario to investigate new forms of consciousness, intelligence, and art in a society radically transformed by technology.

This is a theater to contemplate rather than define, a theater to reflect upon through exploration. Therefore, in this discourse, we do not merely trace the changes in the artistic landscape following technological intrusion; we also delve into a sociological investigation of the altered state of 'being' that this new art represents and the cognitive dissonance it introduces. A theater that is not only mediated but 'mediatized', referencing what Andreas Hepp has outlined with the concept of mediatization;[11] highlighting how society is increasingly saturated and influenced by media logic. This process, particularly relevant for performing arts that use technology, profoundly transforms both the production and reception of artistic works.

In the field of performing arts, the adoption of digital technologies extends and intensifies expressive possibilities, allowing performances to transcend traditional boundaries of space and time. The live streaming of theatrical events to a global audience is an example, as it transforms the performance into a collective experience that surpasses the physical confines of the theater.

The introduction of interactive media and augmented reality in performing arts represents a further intensification of technological interactions, enriching the spectator's experience and offering new modes of storytelling.

These technological tools not only expand the creative possibilities of artists but also modify the expectations and perceptions of the audience.

In this context, the concept of "mediatized dramaturgy" becomes relevant, referring to the design and execution of performative works where digital media are integrated as fundamental components of the narrative and scenic structure. Mediatized dramaturgy does not merely use technology as a scenic support but actively incorporates it into the fabric of the performance, influencing the plot, dialogue, direction, and interaction with the audience. This practice reflects an evolution in theatrical storytelling, where technology becomes a dynamic actor in the story, capable of opening new interpretative and experiential dimensions.

The mediatization of our interest also facilitates the formation of transnational social networks and cultural identities. Performative works that incorporate digital media can serve as cultural bridges, promoting understanding and sharing among diverse audiences and contributing to a culture of connectivity that transcends geographical and cultural boundaries. This process not only changes the volume of artistic production but also profoundly reshapes social structures and functions, accelerating processes such as globalization and individualization in the artistic field. This approach offers a complex and nuanced view of the transformative role of media in the reconfiguration of contemporary artistic practices.

On this, we would like to draw some paths of reflection, which aim not to delineate a map but to follow traces—in an anthropological sense[12]—that can avert the technophilia that sometimes persists in more traditional academic studies. It is now widely accepted that new technologies have radically democratized the creative process, opening new possibilities for artistic expression and collaboration that were unimaginable a few years ago. Through advanced digital tools such as 3D modeling platforms, virtual reality, and global networks, artists can now create, collaborate, and share their works with a vast and diverse audience.

Another significant example is the use of artificial intelligence (AI) and machine learning for artistic creation. These technologies allow artists to explore new creative paths and analyze large amounts of data to generate innovative content. The partnership between human creativity and artificial intelligence not only amplifies artistic capabilities but also enables greater accessibility to creation tools.[13] Additionally, the spread of platforms like Unity or Spatial allows artists to create real-time digital performances, enabling interactions between avatars and artists in different parts of the world, as demonstrated by the performances that will be cited. These developments not only make art more accessible and inclusive but also stimulate diversity in artistic narratives, promoting the representation of traditionally marginalized voices and perspectives. Emerging technologies transform the creative process, making it more democratic and participatory but, above all, more faithful to the image of the human being shared and constructed in contemporary times.

The results of this initial anthropological reconnaissance of the phenomenon underway in the performing arts have hopefully allowed us to overcome some widespread interpretative dichotomies that sometimes prove inadequate for grasping the dynamics at play or for redefining some fundamental concepts for theatrical sciences, but in reality, are more relationally and anthropologically malleable.

Notes

1 R.K. Merton, *Social Theory and Social Structure*, Free Press, New York, 1968, p. 21.
2 D. Haraway, *Staying with the Trouble: Making Kin in the Chthulucene*, Duke University Press, Durham, NC, 2016.
3 C. Geertz, *Local Knowledge. Further Essays in Interpretative Anthropology*, Basic Books, Inc., New York, 1983.
4 L. Floridi (ed.), *The Onlife Manifesto: Being Human in a Hyperconnected Era*, Springer, Cham, 2015.
5 L. Floridi, Metaverse: A Matter of Experience, *Philosophy & Technology*, 35, 73. https://doi.org/10.1007/s13347-022-00568-6; https://link.springer.com/article/10.1007/s13347-022-00568-6#citeas.
6 Roberto Cuppone, a theatrical anthropologist at the University of Genoa, has written about this in his introduction to the contributions of four fundamental theorists and thinkers in the history of theater study in Italy, namely Fabrizio Cruciani, Claudio Meldolesi, Franco Ruffini, and Ferdinando Taviani. Cuppone highlights how these thinkers brought about a Copernican revolution in theater studies, shifting the focus from philology to the physiology of theater. They were the protagonists of a revolution in mainstream academic studies, which, while respecting conventional historiography, challenged its sectorial and critically dramatic approach and sought to break away from the text-centric perspective that had hitherto been prevalent in theater, R. Cuppone, *Pensare il teatro*, Titivillus, Corazzano, 2023.
7 R. Cuppone, *Pensare il teatro*, Titivillus, Corazzano, 2023.
8 R. Krauss, Reinventing the Medium, in *Critical Inquiry*, 25(2), 289–305, 1999.
9 V. Flusser, *Into the Universe of Technical Images,* University of Minnesota Press, Minneapolis, 2011.
10 C. Kattenbelt et al. *Intermediality in Theatre and Performance*, Rodopi, Amsterdam, 2006.
11 N. Couldry, H. Hepp, *The Mediated Construction of Reality*, Polity Press, Cambridge, 2017.
12 C. Geertz, *The Interpretation of Cultures*, Basic Books, New York, 1973.
13 M. Mazzone, A. Elgammal (eds), *Art, Creativity, and Artificial Intelligence, Arts*, 8(1), 26, 2019.

1 Anthropological and Performative Paradoxes on the Contemporary Stage

This chapter delves into the aesthetic and anthropological dimensions of technological hybridization in theater, exploring "paradoxical" paths of analysis. It extends the traditional scope of anthropology, originally focused on humans, to include artificial life, suggesting an evolving field responsive to technological advances like AI and robotics in entertainment. The discussion leverages theatrical anthropology, not conventionally scientific, but empirically pragmatic, to examine the socio-cultural rituals integrating human and non-human elements. It challenges the conventional roles within theater through the lens of myth and anthropology, critically reassessing the transformative impact of technology on modern theatrical practices and their underlying theoretical foundations.

1.1 Deconstructing the Human: An Artistic Paradox

In the contemporary landscape of performing arts, the interaction between the human and the digital is revolutionizing modes of expression and fruition. Artistic performance, traditionally viewed as a bastion of physical presence and human skill, is now undergoing a profound transformation. Digital technologies, with their simulation and interaction capabilities, are redefining the very boundaries of performance, challenging the central role of the human being. In this new context, digital agency[1]—the ability of technologies to autonomously act and influence performative dynamics—assumes an increasingly prominent role. Interactive performances, where sensors, algorithms, and artificial intelligence (AI) become co-protagonists alongside human artists, offer new forms of expression and engagement. As Goffman states, "reality is a performance, and every social interaction is a representation of the self".[2] These technological tools not only amplify creative possibilities but also raise fundamental questions about the meaning of presence and authenticity in art.

The crisis of the status of the human in performance emerges clearly in these experiments. While technology expands the expressive repertoire, on the one hand, it challenges the traditional notion of authorship and artistic

DOI: 10.4324/9781032676937-2

control, on the other. Artists must confront the possibility that their creations may be co-created or modified by non-human entities, questioning the concept of uniqueness and spontaneity that has characterized the performing arts. As Goffman points out, "any break in the current frame can cause a redefinition of the context and a renegotiation of meaning".[3]

Moreover, this evolution raises important ethical and philosophical questions. The relationship between artist and audience is mediated and transformed by the presence of the digital, creating new dynamics of participation and perception. The fusion of real and virtual generates hybrid experiences that can both enrich and confuse the audience, pushing them to reflect on what it means to be spectators in an increasingly technologized world. Goffman asserts that "impression management is central to every social interaction, and new technologies amplify this dynamic".[4] Artists and creators are called to navigate this new territory, exploring the potential and tensions of an art increasingly intertwined with the technologies of our time. As Goffman states, "the play of masks and mirrors is intrinsic to the human condition, and digital technologies only exacerbate this complexity".[5]

To understand the phenomena of technological hybridization from an aesthetic and anthropological perspective, we must now proceed through "paradoxical" paths of analysis. The first is intrinsically linked to the definition of anthropology, closely tied to human beings, although at the end of the 20th century, we witnessed the birth and development of an anthropology of artificial life, which by definition is an analogy of the living, something that begins with a discontinuity or fracture. We now ask the question: why can theatrical anthropology explain the emergence and integration of the most advanced technological discoveries, such as AI and robotics, in the field of entertainment? Theatrical anthropology has never been considered a science, but it is in a broad sense when linked to pragmatic empiricism. If a scientific law derives its validity by demonstrating why fact A follows fact B, a pragmatic law derives its validity by demonstrating that and how fact A follows fact B. The pragmatic law does not state the cause of a relationship but rather highlights its existence and the ways in which acquired facts increase and vary.[6] Therefore, when we speak of the anthropology of artificial life—or a theater made by non-humans—we are simply analyzing an anthropological phenomenon, in the sense of an analysis of a ritual event (strictly socio-cultural) related to humans and non-humans.

The second paradox concerns the "myth", which with its foundational categories is so closely linked to anthropological theater and some archaeological spaces not primarily conceived for theater, and clashes with the actor who has become an artifex, in constant search of synthesis between the categories imposed by the genre and their internal dynamics. "Theatre, anthropology, and myth: it is on these foundations that the modern stage has been resumed and definitively revolutionized, from a perspective that we must consider and critically analyze".[7]

A useful reference for our analysis of the various paradoxes highlighted by contemporary theatrical anthropology is deconstructionism,[8] a concept introduced into Western philosophy by Jacques Derrida. Although the French philosopher avoids any attempt to define deconstruction, it can be described as a close examination of the texts and authors of Western philosophy with the aim of highlighting the implicit assumptions, hidden biases, and latent contradictions of culture and language that humans adopt, often unconsciously. Here, our first anchoring point, there is no catalogable genre of technological performance, no vein in which true stylistic features and aesthetic objectives recur.

Derrida focuses on the idea that meaning is not fixed but rather fluid and ever-evolving. Although he asserts, like Lévi-Strauss, that social, cultural, psychological, and philosophical structures influence or determine the perception and interpretation of the world, he defends the idea that these structures are not fixed or stable but subject to change, ambiguity, and reinterpretation. Similarly, this book aims to emphasize the porosity and protean nature of theatrical art, impacted by the phenomenon of the technologization of artistic and creative language.

Derrida argues that texts and discourses contain hidden differences and oppositions that can be explored to reveal inconsistencies and contradictions. One of the key concepts of deconstruction is *différance*, a term Derrida coined to describe the continuous process of differentiation and deferral of meaning in texts.[9] Through this lens, we can attempt to interpret the paths of artistic creation pursued over the past two decades. In analyzing a technological performance, the concepts of deconstruction can be effectively translated:

- The indeterminacy of meaning allows us to explore how technology alters or amplifies the viewer's interpretations, challenging the traditional perception of the work.
- The critique of meta-narratives can investigate how such performances challenge or dismantle grand narratives or dominant ideologies, proposing new narratives or viewpoints.
- The deconstruction of binaries can be used to analyze how dichotomies like digital/physical, real/virtual, and spectator/participant are blurred or entirely eliminated, reflecting on the fluidity of categories in technology-mediated environments.

Performative paradoxes and a perpetual deconstruction, therefore.

These critical tools provide a more profound understanding of the dynamics and cultural implications of contemporary technological performances.

Let us now consider another valuable philosophical perspective. More than twenty-five years ago, N. Katherine Hayles, in her seminal work,[10] critically and comprehensively examined how the perception of the human has been transformed by cybernetics, computer technology, and AI. Her analysis centers on the concept of the human, challenging traditional notions of human identity grounded in autonomy, bodily integrity, and a clear demarcation between mind and body, human and machine. "The posthuman subject is an amalgam, a collection of heterogeneous components, a material-informational entity whose boundaries undergo continuous construction and reconstruction".[11]

Hayles begins with the evolution of cybernetics, particularly its origins in the 1940s and 1950s, when scientists like Norbert Wiener began developing theories on feedback systems that regulate both machines and living organisms. This idea laid the groundwork for conceiving the mind and body as systems integrated with machines, suggesting a continuity between human and non-human. Hayles' analysis extends beyond technological discussions to address cultural and philosophical implications, critiquing the excessive emphasis on virtualization and the neglect of the body, underscoring the importance of not losing sight of the material and corporeal in the enthusiasm for digital technologies. The technological transition is facilitated by the dismantling of a singular idea of the body/corporeality. In her view, cybernetics is replete with skeuomorphs, design elements that retain the appearance or some characteristics of an older or traditional object but are no longer functional in the new context.

> In the history of cybernetics, skeuomorphs acted as threshold devices, smoothing the transition between one conceptual constellation and another. Homeostasis, a foundational concept during the first wave, functioned during the second wave as a skeuomorph. Although homeostasis remained an important concept in biology, by about 1960 it had ceased to be an initiating premise in cybernetics. Instead, it performed the work of a gesture or an allusion used to authenticate new elements in the emerging constellation of reflexivity. At the same time, it also exerted an inertial pull on the new elements, limiting how radically they could transform the constellation.[12]

The world of replicants, similarly to that of avatars, paradoxically facilitates the viewer in confronting fundamental questions such as the essence of human identity and the concept of self as viewed by society. Technological advancements have indeed enabled the overcoming of numerous practical, cultural, and psychological limits and restrictions. It is no longer our physical body that defines us, raising the question: if we are not our body, what are we? Furthermore, identity varies over time and space, thus revealing itself to be an extremely complex concept to comprehend; one might assert that the nature of a self, aware of its own existence, is yet incapable of fully understanding itself.[13] Following this reasoning, and seeking to apply it specifically in the contemporary performative field, it is useful to refer to the thought of another contemporary philosopher, Brian Massumi, who offers a significant and persuasive analysis of the relationships between issues concerning the body and corporeality and their applications in the contemporary media system.[14] By drawing on Deleuze's concepts of the virtual as potentiality of the possible and affect as an intense experience that determines the subject's departure from itself, Massumi demonstrates how the current audiovisual system is indeed marked by forms and apparatuses that activate physical intensity in the user, providing a novel articulation of the experience of movement and its interrelations with human senses. On these grounds, the philosopher introduces the idea of a continuum between nature and culture that transcends the Enlightenment dichotomy, reconfiguring sensory hierarchies within a relational dimension.

The body of the performer, when hybridized, can, in my opinion, be ascribed to this notion. We are witnessing an integrated humanity that embraces both AI and bodies in a continuous dialogue between material and information. On the contemporary stage, artistic dynamics of significant interest in this regard are manifest. We shall observe how, in order to achieve an ontological understanding of these phenomena, we shall proceed—in the subsequent paragraph—with an analysis of concrete performative examples.

1.2 AI-Driven Performance Inspired by Human

AI, a broad term encompassing various technologies, has now permeated numerous human operational domains, including the once exclusively human domain of theatrical arts. Throughout the history of media arts and subsequently performing studies, artists have approached new technologies with both enthusiasm and skepticism, exploring their potential and limitations, as well as their societal implications.[15] The use of AI in performing arts specifically presents ethical and aesthetic challenges, such as the artist's responsibility in controlling AI and balancing human creativity with automated generation.[16]

However, the integration of these technologies can open new avenues for artistic expression, leading to engaging and innovative performances that challenge our traditional perceptions of art and performance.[17] One of the primary fields of AI application in performing arts is artistic creation. Through machine learning algorithms, it is possible to generate choreographies, stage designs, and even theatrical texts that dynamically interact with human actors.[18] This creation process not only expands artistic possibilities but also introduces new interpretative challenges for performers, who must adapt to an unconventional and constantly evolving choreographic and dramaturgical language.

AI can also transform how audiences interact with artistic performances. Through facial recognition technologies and emotional analysis, it is possible to create immersive experiences that react in real-time to the emotions of the audience. For instance, in some interactive performances, AI sensors and algorithms monitor audience reactions, adjusting the narrative or lighting to enhance the emotional impact of the work. This type of interaction not only makes each performance unique but also fosters deeper and more personal audience engagement.

By using motion recognition algorithms and data analysis, it is possible to obtain a detailed understanding of performative dynamics and the effectiveness of different performance components. This type of analysis can be used both during the development phase of the performance, to optimize choreographies and stage designs, and afterwards, to evaluate the audience's impact and improve future productions.

The use of AI in performing arts also raises significant ethical and aesthetic questions. One major concern is the artist's responsibility in controlling AI. Who is responsible for the creative decisions made by an algorithm? How is human creativity balanced with automated generation? Additionally, the integration of AI can lead to the standardization of artistic practices, reducing expressive diversity. Therefore, it is crucial for artists to maintain a central role in the creative process, using AI as a tool to expand their capabilities rather than as a substitute for human thought. It is now useful to engage with performative examples that find this dialogue and balance between the human and non-human on stage.

Wayne McGregor's Random Dance[19] is distinguished by its innovative approach to contemporary dance, characterized by a seamless integration of cutting-edge technology and human movement, resulting in groundbreaking performances. His work is notable for its frequent collaborations with scientists, software developers, and digital artists, incorporating elements such as motion capture, virtual reality, and AI. These technological integrations not only enhance the visual and sensory experience of his performances but also expand the boundaries of dance perception and experience. Since 1997, when he conceived *53 Bytes*, a performance simultaneously enacted in Berlin and Canada and viewed by audiences via satellite, McGregor has explored the frontiers and potentials of computers, collaborating with animation experts and creators of virtual 3D worlds, as seen in *Sulphur 16* (1998). In this work, dancers appeared diminished by the presence of a virtual giant and numerous digital figures that slithered among them like visitors from another era. Similarly, in *Aeon* (2000), digitally created landscapes transported the dancers to other worlds and dimensions. Technology is explored in different modalities as well, as highlighted in *Nemesis* (2002), where the focus is on the 'machine' as a means of extending human capabilities, embodied here by long metal prostheses attached to the dancers' arms, transforming them into resonant androids on stage.

His choreographic process is highly innovative, involving improvisation and experimentation, which encourages dancers to explore and push their physical limits, creating a distinctive movement vocabulary that is both intricate and fluid. McGregor's interest in science leads him to collaborate with cognitive scientists, biologists, and neurologists, exploring the connections between body, mind, and movement. This interdisciplinary approach reflects a deep understanding of human anatomy and the psychological processes underlying movement.

The performance selected as an exemplar of a 'paradoxical' performance is Wayne McGregor's project *Autobiography*[20] (2017). This work is an abstract meditation on the facets of self, life, and authorship, employing a non-linear narrative that refracts both remembered pasts and speculative futures. In 2017, McGregor collaborated with dancers from his company to devise choreographies derived from old writings, personal memories, and significant pieces of art and music from his life. This creative process resulted in 23 distinct sections of movement, each reflecting the 23 pairs of chromosomes in the human genome. These choreographic sequences were integrated into a sophisticated algorithm based on McGregor's genetic code. This algorithm, developed in collaboration with Nick Rothwell, utilizes coding principles and AI to dynamically generate the sequence of the performance. For each performance, the algorithm randomly selects a segment of code from McGregor's genome, which dictates the order and specific content presented to the audience, framed by predetermined beginning and end points.

By superimposing McGregor's choreographic signature onto his personal memories and genetic data, *Autobiography* continually reinvents itself, unfolding in a unique configuration for each performance. This integration of AI and genetic coding not only innovates the narrative structure but also mirrors the dynamic and ever-evolving process of life itself. The paradoxical nature of this performance is particularly striking because it originates from the organic and biological data of McGregor's genetic code and, through an artificial system of synthetic programming, generates an ever-evolving artistic product. The use of an algorithm to dictate the choreographic sequence symbolizes a fusion between human organicity and artificial mechanisms, creating a dynamic performance that evolves in a manner akin to living human cells. This interplay challenges traditional boundaries between the organic and the synthetic, highlighting a complex dialogue wherein biology converges with technology.

The notion that an algorithm can not only interpret but also dictate artistic expression introduces a layer of irony to the project. It suggests that the deterministic elements of our biology can be intertwined with the stochastic nature of computer algorithms to create something unpredictably human and vibrant with life. The involvement of the computer in the creation of a narrative framework or theatrical dramaturgy is not an innovation exclusive to our contemporary period. As early as 1960, the linguist Joseph E. Grimes developed an algorithm capable of generating stories. More than thirty years later, MIT Media Laboratory scholars Claudio Pinhanez and Aaron Bobick devised performances involving interaction between a human character and one portrayed by a computer.[21]

This raises intriguing questions about the role of AI in art. "Algorithms play a crucial role in creative generation, capable of following specific rules or exploring creative spaces more freely, producing outputs that can surprise and inspire human creators".[22]

Algorithms can learn and replicate artistic styles, creating works that reflect specific aesthetic trends or combine different styles in original ways, pushing humans to consider unexplored and unimaginable canons. Consider, for example, the use of Generative Adversarial Networks (GANs) to create art, trained on vast datasets of images to generate new works that can be indistinguishable from those created by human artists.[23]

> If the interplay between man and machine had to become more "à la pair", it was necessary to lead the AI to choose, to identify what it saw. With this ability to choose – choose one identity, choose what something is – AI art really starts to say something different about art and also about art's meaning beyond the merely artistic sphere. In fact, with this step the creation process becomes a real dialogue with an 'other' that is very similar to us: it can make mistakes; it can fail. And it has to be listened to. [...] The interplay, within AI art, is never autocratic, but is always connected to the capacity for listening to another "tongue", another way of reasoning. This is the old message that AI art brings to light with a new clarity in the artistic sphere. It is necessary to go back to the perceptive sphere, and to the moment of listening, in order to find an image of the world that can be shared by me and my many "others", humans or machines.[24]

In this way, *Autobiography*[25] serves as a reflective mirror and a forward-looking probe, questioning and expanding our definitions of life and creativity, positing a future where artificial and natural syntheses yield new forms of artistic expression. This project is a testament to the evolving narrative of what it means to be both a creator and a creation in the age of advanced technology.

According to Lev Manovich, the assumptions that art, compared to any other field of human activity, embodies creativity the best and that art is the best expression of human uniqueness, lead to the following apparently logical conclusion: the best test of the progress of AI (or of the most advanced robotics) is if it can generate new art. Here we encounter another fascinating paradox.

This paradox concerns dance, an art of the body that allows us to see and communicate through it. Yet, we have seen how in the last decade, the desire to overcome certain bodily limits or characteristics, or simply the drive to integrate technological means that now constitute the interface or tool for many daily actions, has led the art of choreography toward a process of corporeal dematerialization.[26] It is difficult to write about 'performance' as much as it is to realize ideas and thoughts, and as such abstractions, on stage, because translating an idea into a work of art is always indirect, representative, and constitutes a map rather than a territory. And the stage is always there, a concrete place first and foremost, a territory that constrains but also accommodates, where a spectator witnesses, observes, in all the physicality that this aesthetic experience entails.

Is it more appropriate to discuss originality or creativity when addressing robotic or dematerialized authors and directors? While the influence of the programmer behind a robot is evident, AI raises different issues, particularly concerning the identification of the object-work versus its author. Copyright law necessitates an active role in the creative process to qualify as an author, a principle that AI challenges, as its creations are not directly attributable to human input, thereby severing any causal link between the artist's concept and the AI's creation. Consequently, the question of authorship of AI-generated works remains unresolved, provoking debate on whether recognition should be attributed to the programmer, the artist/user, the AI itself, or to no one, thus relegating such works to the public domain. Furthermore, even AI-generated works where the artist exerts substantial control over the creative process—thereby ensuring authorial attribution—are fraught with ethical, racial, and content manipulation complexities.

Technophiles assert that AI will soon surpass the capabilities of the most gifted human minds. However, examining the artistic and performative experiments conducted thus far, involving AI as author or performer, reveals that inductive reasoning or the analysis of datasets to predict possible outcomes cannot compare to the conjectures informed by context and experience employed by human interpreters. Nevertheless, the interaction between humans and machines, between the organic and the synthetic, gives rise to a poetics and aesthetics that are increasingly recognizable to audiences. For instance, the recent experiment in which the performer and researcher Valencia James participated,[27] *AI_am* (2013), serves as a pertinent example. One of the critical roles that AI fulfills in enhancing dance choreography pertains to the analysis of movements.

> The project focuses on creating an improvised duet between an AI and a human dancer. The -, projected on stage, observes and learns the movements of the dancer, but also extends them with its own variations. Throughout the performance, the distinction between "teacher" and "student" gradually blurs, as the human dancer begins to find inspiration in the avatar's novel movements.[28]

The interdisciplinary initiative *AI_am* delves into the symbiotic relationship between AI and contemporary dance, aiming to enrich and advance both fields reciprocally. The project unites specialists from the realms of dance, cognitive sciences, graphic design, and programming. At the core of the project is the creation of an improvised duet between an AI and a human dancer. Presented on stage, the avatar observes and assimilates the dancer's movements, adding its interpretations. As the performance unfolds, the boundaries between "teacher" and "student" gradually blur, with the human dancer drawing inspiration from the avatar's innovative moves. By integrating advanced machine learning methods into the world of dance and orchestrating a distinctive convergence of art and science, the project aims to reinvent the essence and potential of the performing arts (Figure 1.1).

Traditionally, choreographers would spend hours observing and analyzing dancers' movements, seeking to understand the nuances and complexities of their performances. With AI, this process becomes much more efficient and precise. AI algorithms can analyze video recordings of dancers and break down their movements into data points, allowing choreographers to gain a deeper understanding of the underlying dynamics of the movements. This not only saves time but also provides valuable insights that can influence the creation of new choreographies.

Another way in which AI enhances dance choreography is through the generation of movement sequences. AI algorithms can be trained on vast amounts of dance recordings, simultaneously learning the patterns and styles of different choreographers. With this knowledge, AI can generate new movement sequences inspired by the work of these choreographers. This opens up a world of possibilities for choreographers, enabling them to explore new ideas and styles of movement that they might not have considered previously. It also provides choreographers with a way to collaborate with AI, creating a unique fusion of human creativity and AI.

The introduction of AI in the realm of choreographic creation challenges those "techniques of the body" fundamental to live performance, as defined by Marcel Mauss. He referred to styles, skills, and bodily formulas that, although appearing natural, are learned from the environment, culture, and society, and thus are implicated in the history of technology and media themselves.[29] Social learning and cultural variability are disrupted and transcended by the creative and generative power of AI systems. This inevitably risks a split or multiplication of the self-body and the social identity assumed by humans.

Valencia James explains:

> This has really expanded my idea of what dance could be. We didn't give it a real-world physics, so it could make movements that wouldn't be humanly possible, but I found that it was really inspiring and generative to explore what it means to move in the style of impossibility.[30]

The idea of an AI learning from a dancer is fascinating and represents a field where technology intertwines with art innovatively. In this scenario, an AI system would be programmed to observe and analyze the movements of a human dancer. Employing machine learning algorithms and computer vision

Figure 1.1 *AI am here*, performance still. 2017. *AI_am*, project by Valencia James, Alexander Berman, Gábor Papp, Gáspár Hajdu and Botond Bognár. Photo credit: György Jókúti.

techniques, the AI can understand patterns, expressions, and dynamics of human movement. The intriguing aspect of this interaction is the potential for a dynamic relationship between the AI and the human dancer. While the AI learns from the dancer's movements, it influences their style or suggests new creative directions. This type of hybrid collaboration between human and machine could lead to surprising and stimulating artistic results, exploring the boundaries between technology and human creativity. It is an artistic relationship comparable to quantum entanglement, a correlation between particles that remain interconnected even over great distances, influencing each other instantaneously. This concept can be used to describe the migration of the performer's body into the metaverse, where physical and virtual presence intertwine, creating an immediate and continuous connection between real movement and the digital avatar, in this case partially autonomous thanks to the use of AI.[31]

This relationship is so close that the very concept of author/choreographer is undermined, raising intriguing philosophical and artistic questions regarding the role of AI in artistic creation. Traditionally, the author is associated with the human ability to conceive, design, and shape a work of art, granting the creator a certain degree of control and responsibility over the final outcome. However, with the advent of advanced AI technologies, this concept is challenged as AI can be programmed to generate works of art, including texts, music, images, and more.

The use of AI in the arts raises questions about the true authorship of works produced in collaboration with an artificial system. If a work is primarily generated by AI, to what extent does the authorship belong to the programmer who created the algorithm, or to the system itself? And if AI is used as a creative tool by a human artist, to what extent is the work still considered theirs and not the AI's? Moreover, AI can challenge the very concept of artistic originality. Since AI can analyze vast datasets of existing works and generate new content based on such analyses, it can be difficult to determine whether a work produced by AI is truly original or merely a repetition of existing elements. These issues present interesting challenges for the art world, including the need to reconsider the concept of authorship and to find a balance between the innovation offered by AI and the respect for human creativity.

> According to ChatGPT, there are countless ways artificial intelligence can be useful to dance artists: Need a brainstorming partner? Help planning rehearsals? A tool for generating new movement? A way of documenting your work? Look no further than the buzzy chatbot technology, it told me when I asked. But don't worry: The chatbot also said that "while ChatGPT can be a valuable tool for choreographers, it should not replace the artistic intuition and expertise that come from years of training and experience."[32]

A significant aspect of the concept of authorship undermined by AI also concerns the nature of inspiration and originality. While AI can be programmed to imitate the style of a particular artist or to produce works that closely resemble existing ones, some may raise doubts about the true creativity and uniqueness of such works. Human creativity often emerges from a complex process of personal experiences, emotions, and reflections, which can be difficult for an algorithm to replicate or imitate. That said, the use of AI in the arts can influence the value and perception of artistic authenticity. Artworks are often considered valuable not only for their aesthetic value but also for their cultural and historical significance, as well as their connection to human experience. The proliferation of AI in the arts also raises ethical and social issues regarding the automation of creative work and its potential impact on the employment of human artists. While AI can be a powerful tool to enhance productivity and stimulate new forms of artistic expression, it can also raise concerns about the replacement or devaluation of human labor in the creative field.

1.3 Collaborative Creativity in the Digital Age: Exploring AI as a Co-Author

Invention and illusion are two concepts that can instinctively be associated with new technologies; when applied in the context of contemporary performance, they take on added significance. This work explores these concepts by presenting technological devices within the performative context not as passive entities, but as active agents, creative subjects capable of generating new things and illusions. At first glance, this may appear paradoxical. However, the artistic domain, which has always been closely tied to humans, initially welcomed the machine as a support and technical device, and subsequently integrated it as a creative partner. Humans are highly adaptive beings capable of surviving and thriving even in unfavorable conditions by creatively exploiting the opportunities and limitations imposed by their environment. Advanced technologies, in particular, necessitate specific forms of adaptation based on the primacy of perception and the construction of new models of self-exhibition that transcend imitative forms, pushing toward the dematerialization and sometimes the depersonalization of the performative body.

AI, virtual environments, and new formats combine the ancient tradition of storytelling with contemporary narrative forms rooted in digital technology. This analysis addresses this timely phenomenon through several performative examples where AI assumes the role of author, playwright, or choreographer. We begin by examining the new forms of authority linked to Chat GPT-3.[33] Increasingly, artists—motivated by challenge or curiosity—entrust aspects of dramaturgical creation to generative language systems.

These artistic explorations highlight the shifting paradigms in performance arts, moving away from an anthropocentric perspective toward a more integrated and symbiotic relationship with technology. The implications of these shifts extend beyond the realm of art, prompting reconsiderations of creativity, authorship, and the very essence of human-machine interaction in contemporary society.

Many now associate the myth of AI with the myth of Prometheus, the figure who stole fire, symbolizing life, from Zeus and used it to cook for all humanity. Naturally angered, Zeus descended to Earth to claim his share of every animal cooked by humans, but Prometheus deceived him, convincing him to choose only the entrails and offal. The deity reacted as gods always do when someone attempts to usurp their power and authority: he punished Prometheus by blinding him and binding him to a rock peak, where an eagle devoured his liver every night. Prometheus's liver would regenerate, and each day the eagle would return to consume it anew. This story undoubtedly represents human ambition to constantly develop new powers and is a testament to the seemingly inexhaustible innate creative spirit of human beings. It is also a tale of hubris: Prometheus could have kept the fire and spared his liver by offering Zeus the choicest cuts. Our deep desire for genuine AI draws inspiration from this myth: we wish to steal fire from the gods despite the potential for terrible consequences. Prometheus was a punished hero; it is no coincidence that Mary Shelley titled her masterpiece "Frankenstein, or the Modern Prometheus". The story demonstrates how creating intelligence through science and technology, infusing life into inanimate matter, is a dangerous attempt by humans to approach the divine and assume its role.[34]

The "invention of the machine" seems like a play on words, but it refers to a contemporary process in which the superintelligent machine invents rather than being invented, as evident in the performance *Una Isla* (2023) by the company Agrupación Señor Serrano.[35] As with any mythology, the creator's fear of being surpassed and annihilated by their creation has led to the dilemma of coexistence between humans and AI. The fear of 'the other, of the unknown', of a superior entity.[36]

> The creation process of *Una isla* has been carried out in dialogue with a series of artificial intelligences that generate text, images and music. The work methodology has consisted in the fact that the artists in charge of each one of the areas of creation entered into a creative dialogue with said intelligences, developing a conversation that has been fed with external inputs and has continuously fed back. Why to use AIs in a creation process? Because it opens up new and different possibilities of thinking and imagining on stage. Because it means giving the show an unforeseen and tangential component that we could never achieve with just human collaborators or with the different creation tools that we have. It is not about making an exhibition of the potentialities of the different AI, but about using them for the benefit of the creation and that they serve as dramaturgical and discursive reinforcement for the concept and poetics that we want to develop.[37]

The stage appears empty, and as darkness envelops the room, a conversation between two interlocutors is projected on the back wall: the white questions come from a human interlocutor—a member of the Agrupación Señor Serrano company—while the yellow responses are generated live by an AI. The script seems to spontaneously generate on stage, but it is actually the result of two years of study and hybrid dramaturgy. In the correspondence, the question "Who are you?" recurs, but the AI's response varies each time: "I am a handful of words in a universe of silence", "I am a yellow owl", "I am the other, but we are all the other to someone". The 'Yellow Converser' can generate endless responses until one appears that the 'White Converser' finds satisfactory or fitting; thus, in the unlimited openness to possibility, the impossibility of a true dialogue manifests (Figure 1.2).

The staging soon becomes clear: on stage, the propositions put forth in the AI chat materialize in flesh and bone, like the insights of a director-demiurge freed from the constraints of logic. Meanwhile, through video and imagery, the island's world is constructed, a layered universe with a history, a past, a present, and an endangered future.

According to Hans-Thies Lehmann, the omnipresence of media in daily life since the 1970s is the driving force behind the development of post-dramatic theater.[38] This new form of theater is more fragmentary than linear, more performative than representational, and the text is merely one element among others. Abolishing the dominance of the text has indeed led to a revaluation of

Figure 1.2 Una Isla. Dancer Carlota Grau Bagès. Photo by Agrupación Señor Serrano.

other theatrical elements. Media on stage, like other technological elements, can thus have intrinsic significance that is not necessarily linked to a narrative dimension. This presents new challenges for spectators: correctly interpreting theatrical signs for which they cannot refer to their previous experience.[39]

Through a mechanism of accumulation and amplification of nonsense, the performance positions itself between a quasi-real dystopia akin to Black Mirror and a new and successful theater of the absurd.[40] Here, we see clearly how the crucial role of AI lies primarily in its function as a new aesthetic indicator, where aesthetics is understood as a concept more closely tied to creativity in a broad sense, encompassing the study of social categories and the judgment of taste, which play a role in processes of knowledge and awareness.

Thus, AI, this totem and taboo that has been stirring the theatrical world for almost a decade, takes the stage and becomes the director, the puppeteer of performers who embody what is evoked by the words of an absent entity, but one that we know to be a superintelligent system.

> Replies from the AI respondent are instant. There is no wish to stop and rethink, to pause and start again. Slow culture has been replaced with quick and easy answers – instant gratification. This is what AI offers. The answers may appear improbable, but they are sold with a line of confidence and repeated allegations of impartiality. As Serrano and Palacios scratch at the surface of what impartiality means, it is all too clear that the language models from AI that they are engaging with have a repository that is anything but impartial. Answers on the story Serrano and Palacios should

> construct are very much driven by Western classical tropes and mythical narratives, variations on *Robinson Crusoe* with the AI respondent unable to interrogate the ideological implications of what they are proposing.[41]

The production includes a series of reflections on the "other"—both in the questions Serrano and Palacios pose to the AI interlocutor and in the variations of movements provided by the physical performers and in the holographic videos. The work explores how new technologies in art are dismantling an anthropocentric view of humanity, a crucial aspect in the anthropology of artificial life.[42] How we define ourselves, what identity means, and what consciousness is possible in AI are effectively probed in the brief on-screen conversations.[43] It is never entirely clear who 'the other' is that is consistently referred to. Indeed, the production seems to suggest that each of us is 'an other' to those we interact with.

In just over an hour on stage, the myth of conflict is presented in a modern key. The myth of the conflict between humans and AI manifests itself through a series of misconceptions similar to those characterizing conflicts between ethnic groups. Just as it is erroneously assumed that ethnic identities are ancient and immutable, that these identities provide motives for persecution and murder, and that ethnic diversity inevitably leads to violence, the human-AI conflict is fueled by similar myths.[44]

AI is often viewed as an autonomous and static entity that cannot evolve or be harmoniously integrated with human activities. However, AI is a product of human programming and adaptation capabilities, and its identity is continuously evolving in response to human needs and social contexts.[45] It is often believed that AI is inherently oriented to supplant or destroy humanity, fostering fears of replacement and loss of control. This mirrors the erroneous belief that ethnic differences are sufficient to justify persecution. In reality, AI, as a tool created by humans, operates within the limits and directives set by its creators and users, aiming to improve human quality of life.[46]

The assumption that the introduction of AI in various sectors will inevitably lead to conflicts and mass unemployment is comparable to the belief that ethnic diversity inevitably leads to violence. However, studies show that AI can be integrated complementarily with human capabilities, creating new types of jobs and improving productivity,[47] though it is unlikely that AI will become more intelligent than its creator.[48]

The use of technologies such as AI and holography challenges human centrality, shifting the focus to a more symbiotic interaction between humans and machines.[49] It is in the power of listening—of bodies tuned to each other to avoid collision, of humans interacting in nuanced ways—that our humanity resides. AI does not yet allow this, at least not fully. The responses generated by the AI interlocutor may be intelligent, but they never seem profound or truly human.

An ironic reflection on the media manipulation of reality is masterfully represented by the Spanish collective. We agree with theorist Lev Manovich when he states,

> The hypothesis that art, unlike any other human activity, best embodies creativity, and also that art is the best expression of human uniqueness, leads to the following seemingly logical conclusion: the best test of AI progress is its ability to generate (new) art.[50]

Here, we encounter a fascinating paradox.

Not all artistic practices that deal with technology or use living beings need to be considered "post-human material". In a certain sense, creating post-human art also implies being (a set of) post-human subjectivities, consciously understanding and exercising certain ethics that include not only the suppression of anthropocentric conceptions in relationships with the living, the semi-living, and the inorganic but also the ability to conceive new methodologies of thought and practice.[51] *So could we Understand the human by using AI or robots, reaching a new theater?*

> The post-human does not necessitate the obsolescence of the human; it does not represent an evolution or devolution of the human. Rather it participates in redistributions of difference and identity (…) the post-human does not reduce difference-from-others to difference-from-self, but rather emerges in the pattern of resonance and interference between the two.[52]

In the realm of theater, both the perception of a world that exists independently of our will and presents itself to our gaze as something inevitable and objective, and the awareness that the spectator experiences in realizing they are a sensitive subject of a perceptual phenomenon during the performance, are stimulated.[53] The subjectivity in appropriating the object "imparts an emotional nuance to these contents, in relation to an experience, to attentive conditions, as well as to the emotional properties to which it is connected, by virtue of its own action and related empathy, the perceptual activity itself".[54]

Although the spectator's engagement is fundamental to the performance, the recognition, processing, and experience of an action and emotion are renegotiated in the hybrid performance described, leading to a type of engagement as a "standalone act". After a period of discovery and innovation, followed by evolution and experimentation, we are now faced with contamination, but this should not lead to forgetting or erasing, nor overwhelming, the foundational ritual that constitutes theater.

Let us now proceed with another artistic example where AI plays a co-protagonist role. In *dSimon*[55] (2021), Geneva-based performer and director Simon Senn collaborates with computer engineer Tammara Leites[56] to "train" a GPT-3 AI system—the most well-known AI software in the world—to become a writer. The duo decides to attribute human characteristics to the 'superintelligent' system, belonging to the artist's personality, and to this end, dSimon, the synthetic writer twin, is given access to all of Senn's personal data (diaries, emails, social profiles, etc.).

On stage begins this lecture-performance in which Tammara and Simon recount their surprising encounter with this digital entity, which has become autonomous, similar to a contemporary dematerialized Frankenstein. These lecture-performances, considering classical aesthetic foundations, in post-modern theater, see the collapse of the art/life, fiction/reality distinction. In the 2000s, a new aesthetic composed of interactions and transformations emerges, a possibility to make seemingly distant elements like science, technology, and nature cooperate and interact. As Latour states: "The lecture-performance is a way to stage ideas, making them more accessible and vivid for the audience".[57]

Audience members are asked, in turn, to say a random word or describe the outfit they are wearing, providing the starting point for the creation of a text by dSimon. This experiment sometimes provokes laughter and, at times, reveals its limitations, deliberately highlighted by the artists on stage, as it often produces too much nonsense that alienates the audience. This demonstrates the machine's fallibility compared to humans and leads us to some questions: why does dSimon seem human? Is it truly autonomous?[58] Can we consider it an original writer or possess creativity equal to that of humans? We witness, more as observers than as an audience, the birth of a 'digital human performer' (Figure 1.3).[59]

Figure 1.3 dSimon (2021). Photo © Niels Ackermann/Lundi13.

Artificial intelligence, which now relies on artificial neurons and deep learning, progressively erases the boundaries that remind us that a conversational agent, for example, is merely code. This agent can be embodied by an avatar, a particularly advanced human-computer representation, increasing our confusion. These digital humans can produce a full range of human languages (both spoken and body language), supported by AI, to interact with people. As successors to chatbots, they appear emotionally intelligent. Their resemblance to human beings is striking and allows the user to have a faithful and realistic representation of their interlocutor.[60]

> Digital humans are human-like virtual AI characters designed to be interacted with by people. They are, for all intents and purposes, the chat bot's successor: sharper, emotionally intelligent, and expressive AI entities embodied in human-like avatars, who, in their often striking resemblance to real people, both in behavior and appearance, give users a face, a personality - a *person* - to relate to. Thus, digital people effectively become "someone," as opposed to some*thing*; their likeness to real humans renders them radically more engaging to users than chatbots. While not even the cleverest engineering could make these "beings" flesh and blood, they represent a significant step toward an AI assistant [...] capable of connecting with people in ways far more profound than those a bodiless chatbot can muster.[61]

In this attempt to humanize the machine, to generate empathy with the audience, the excessive "freedom", also called autonomy, has proven to be the primary challenge for the synthetic author. This performance by Senn—but also in one that will follow—brings us back to what Peggy Phelan expressed in her *Unmarked: The Politics of Performance*,[62] to her exploration of the concept of "unmarked" as a mode of resistance to traditional cultural binaries, such as self/other, man/woman, presence/absence. These themes are reflected in *dSimon*'s performance, where the dichotomy between human and AI is continuously interrogated and subverted. Phelan's analysis of the ephemeral and the unrepresented highlights how stage presence is in constant negotiation between visible and invisible, between human and technological in our case. Senn's work stages a critique of the stability of identity categories and a celebration of the indeterminate and fluid, central aspects of Phelan's work. AI, now on stage like robots, has shown itself capable of 'human' qualities: it is certainly autonomous, partially creative, but not at all conscious.

Singularity is still distant,[63] but in general, we are witnessing a paradigm shift even in the field of art, no longer generated from a merely anthropocentric perspective. The fundamental role of AI lies in its function as an aesthetic indicator, where by aesthetics we mean a concept more closely related to creativity in a broad sense and which includes the study of social categories of taste judgment that play a role in processes of knowledge and awareness.[64]

Notes

1 T. Wheeler, *Artificial Intelligence Is Another Reason for a New Digital Agency*, in Brookings (blog), 28th April 2023, https://www.brookings.edu/blog/techtank/2023/04/28/artificial-intelligence-is-another-reason-for-a-new-digital-agency/, [accessed 25 May 2024].
2 E. Goffman, *The Presentation of Self in Everyday Life*, Anchor Books, London, 1959.
3 E. Goffmann, *Frame Analysis: An Essay on the Organization of Experience*, Harvard University Press, Cambridge, MA, 1974.
4 E. Goffman, *The Presentation of Self in Everyday Life*, Anchor Books, London, 1959.
5 E. Goffmann, *Forms of Talk (Conduct and Communication)*, University of Pennsylvania Press, 1981.
6 L. Allegri, *Prima lezione sul teatro*, Edizioni Laterza, Bari-Roma, 2012.
7 L. Allegri, *Prima lezione sul teatro*, Edizioni Laterza, Bari-Roma, 2012.
8 See J. Derrida, *De la grammatologie*, Éditions de Minuit, Paris, 1967.
9 J. Derrida, *De la grammatologie*, Éditions de Minuit, Paris, 1967.
10 N. Katherine Hayles, *How We Became Posthuman: Virtual Bodies in Cybernetics, Literature, and Informatics*, The University of Chicago Press, Chicago, IL and London, 1999.
11 N. Katherine Hayles, *How We Became Posthuman: Virtual Bodies in Cybernetics, Literature, and Informatics*, The University of Chicago Press, Chicago and London, 1999, p. 4.
12 N. Katherine Hayles, *How We Became Posthuman: Virtual Bodies in Cybernetics, Literature, and Informatics*, The University of Chicago Press, Chicago and London, 1999, p. 18.
13 See H. Ishiguro, *How Human Is Human?The View from Robotics Research*, Japan Publishing Industry Foundation for Culture, Chiyoda, Tokyo, 2020, pp. 48–54.
14 See B. Massumi, *Parables for the Virtual. Movement, Affect, Sensation*, Duke University Press, Durham, NC and London, 2002.
15 See L. Jones, *Exploring New Frontiers in Media Arts*, Oxford University Press, New York, 2020; and A. Smith, *Technological Innovations and Their Societal Impacts*, Cambridge University Press, Cambridge, 2019.
16 See D. Brown, J. Harris, *Ethics in the Age of Artificial Intelligence*, Routledge, London, 2021.
17 See S.Gill, A. Dorsen, The Work of Art in the Age of Digital Commodification: The Digital Political Economy of the Performing Arts, in *TDR: The Drama Review* 68(1), 19–50, 2024, doi: 10.1017/S1054204323000618.
18 R. Wingström, J. Hautala, R. Lundman, Redefining Creativity in the Era of AI? Perspectives of Computer Scientists and New Media Artists, in *Creativity Research Journal* 36(2), 177–193, 2022, doi: 10.1080/10400419.2022.2107850.
19 https://waynemcgregor.com/, [accessed 20 March 2024].
20 https://waynemcgregor.com/productions/autobiography/, [accessed 20 March 2024].
21 Claudio Pinhanez and Aaron Bobick conducted a significant performance experiment entitled "It/I". This project is a two-character play in which the human character "I" interacts with and is provoked by an autonomous computerized character called "It". The project was implemented and presented at the MIT Media Laboratory in

January 1998. "It/I" was one of the first attempts to integrate a computer-controlled character into a theatrical performance. Using computer vision, the system was able to recognize and respond to the actions of the human character in real time, creating a unique interactive experience. The project demonstrated how autonomous characters can add a new dimension to theater by allowing the audience to interact directly with the characters on stage after the performance, re-invoking the story of the play. See https://www.media.mit.edu/publications/iti-a-theater-play-featuring-an-autonomous-computer-graphics-character-2/, [accessed 20 March 2024].

22 J. McCormack, M. d'Inverno, *Computers and Creativity*, Springer, Berlin and Heidelberg, 2012.

23 A. Elgammal, *Art and Artificial Intelligence*, Columbia University Press, New York, 2019.

24 A. Barale, Who Inspires Who? Aesthetics in Front of AI Art, in *Philosophical Inquiries* 9(2), 195–220, 2021, doi 10.4454/philinq.v9i2.

25 Thinking about his choreographic work-an integral part of his unique archive-McGregor was intrigued by how artificial intelligence (AI) could model a way of looking at his past and current practice and then propose future possibilities for movement: a system of 'life writing' in relation to choreography. In 2019, this led him to a collaboration with Google Arts & Culture Lab. Taking the entire catalog of his past work as a dataset, the lab used machine-learning technology to create an AI-powered tool, Living Archive, that can suggest new original movement phrases based on McGregor's existing choreography and material he created on individual dancers. See https://waynemcgregor.com/about/wayne-mcgregor/, [accessed 20 March 2024].

26 B. Hookway, *Interface*, MIT Press, Cambridge and London, 2014.

27 Valencia James is a Barbadian freelance performer, maker and researcher interested in the intersection between dance, theater, technology, and activism. She believes in the responsibility of artists to reflect socio-political issues and in the power of the arts to inspire change. The team project: Valencia James – Lead artist, Research, choreography and performance. Botond Bognár – Concept and research. Alexander Berman – AI research and development. Gábor Papp – Creative coding, rendering. Gáspár Hajdu – Motion capture, rendering, architectural design. https://valenciajames.com/about/, [accessed 10 October 2023].

28 https://valenciajames.com/projects/ai_am/, [accessed 10 October 2023].

29 See M. Mauss, *Les technique du corps*, Paris, Payot, 1936 (2021) and M. Mauss , *Manuel d'ethnographie* (1926) Éditions sociales, Paris, Payot, 1936 (1967).

30 https://www.uncsa.edu/kenan/art-restart/valencia-james.aspx, [accessed 10 October 2023].

31 A.D. Aczel, *Entanglement: The Greatest Mystery in Physics*, Four Walls Eight Windows, New York, 2001.

32 L. Wingenroth, *How are Dance Artists Using AI and What Could the Technology Mean for the Industry*? in Dance Magazine, 24th July 2023, https://www.dancemagazine.com/how-dancers-use-ai/, [accessed 10 October 2023].

33 ChatGPT-3, developed by OpenAI, is a state-of-the-art language model known for its ability to generate human-like text based on a given prompt. It has been widely used for various applications, including content creation, customer support, and conversational agents. The latest version, ChatGPT-4, builds on these capabilities with improved understanding and generation of nuanced responses, enhanced context retention, and broader knowledge base. For more information, visit https://www.openai.com.

34 E.J. Larson, *The Myth of Artificial Intelligence. Why Computers Can't Think The Way We Do*, The Berknap Press of Harvard University Press, Cambridge, MA, 2021.

35 https://www.srserrano.com/it/.

36 https://www.srserrano.com/it/una-isla-ita/.
37 https://www.srserrano.com/it/una-isla-ita/.
38 See H.-T. Lehmann, *Postdramatic Theatre*, Routledge, 2006.
39 S. Hagemann, I. Pluta (eds), *Quels rôles pour le spectateur à l'ère numérique?*, Épistémé, 2023, p. 55, open access e-book available https://www.epflpress.org/produit/1480/9782889155699/quels-roles-pour-le-spectateur-a-l-ere-numerique.
40 M. Degado, *Floating Islands of AI: Agrupaciòn Serrano's Una Isla/The Island*, 27th October 2023, https://thetheatretimes.com/floating-islands-of-ai-agrupacion-senor-serranos-la-isla-the-island-in-madrid/.
41 M. Degado, *Floating Islands of AI: Agrupaciòn Serrano's Una Isla/The Island*, 27th October 2023, https://thetheatretimes.com/floating-islands-of-ai-agrupacion-senor-serranos-la-isla-the-island-in-madrid/.
42 See N. Katherine Hayles, *How We Became Posthuman: Virtual Bodies in Cybernetics, Literature, and Informatics*, The University of Chicago Press, Chicago and London, 1999.
43 In the two selected performances, we find alignment with what Steve Dixon outlines in his recent volume *Cybernetic-Existentialism: Freedom, Systems, and Being-for-Others*. Artificial intelligence (AI) interacts with human performers, challenging the boundaries between human and machine, as well as the relationship between author and creation. Dixon discusses how technological systems influence our self-understanding and existential freedom, See S. Dixon, *Cybernetic-Existentialism: Freedom, Systems, and Being-for-Others*, Cambridge, MA, 2020, pp. 68–107.
44 See J.R.Bowen, Il mito del conflitto etnico globale, in F. Dei (ed.), *Antropologia della violenza*, Meltemi, Roma, 2005.
45 See E. Brynjolfsson, A. McAfee, *The Second Machine Age: Work, Progress, and Prosperity in a Time of Brilliant Technologies*, W.W. Norton & Company, New York, 2014.
46 Y.N. Harari, *Homo Deus: A Brief History of Tomorrow*, Harvill Secker, London, 2017.
47 D.H. Autor, Why Are There Still So Many Jobs? The History and Future of Workplace Automation, in *Journal of Economic Perspectives* 29(3), 3–30, 2015.
48 J.R.Bowen, Il mito del conflitto etnico globale, in F. Dei (ed.), *Antropologia della violenza*, Meltemi, Roma, 125–144, 2005.
49 See E. Brynjolfsson, A. McAfee, *The Second Machine Age: Work, Progress, and Prosperity in a Time of Brilliant Technologies*, W.W. Norton & Company, New York, 2014.
50 L. Manovich, *The Language of New Media*, The MIT Press, Cambridge, 2001.
51 See N. Katherine Hayles, *How We Became Posthuman: Virtual Bodies in Cybernetics, Literature, and Informatics*, The University of Chicago Press, Chicago, IL and London, 1999.
52 J.M. Halberstam, I. Livingston (eds), *Posthuman Bodies*, Indiana University Press, Bloomington, 1995, p. 10.
53 E. Fischer-Lichte, *The Transformative Power of Performance: A New Aesthetics*, Routledge, New York, 2008.
54 E. Fischer-Lichte, *The Transformative Power of Performance: A New Aesthetics*, Routledge, New York, 2008.
55 https://www.simonsenn.com/dsimon/.
56 Born and raised in Uruguay and now living in Geneva, Tammara Leites has always been passionate about technology and society's relationship with it. With a formal background in programming, graphic design and visual communication, she decided to undertake a Master's degree in Media Design at the Haute école d'art et de design (HEAD) in Geneva in order to conceive projects that allow both her

creativity and interests to converge. Her work reflects upon what it means to be a connected human every day. Simon Senn was born in 1986 and lives in Geneva. He has completed a Bachelor of Fine Arts at HEAD in Geneva and a Master's at Goldsmiths College in London. At first glance, his work seems to suggest that he is a socially committed artist speaking out against a certain type of injustice. However, his work sometimes reveals a more ambiguous approach, exploring paradoxes rather than articulating directed criticism.

57 B. Latour, *Facing Gaia: Eight Lectures on the New Climatic Regime*, John Wiley & Sons, Hoboken, NJ, 2017, p. 15.

58 E. Fuoco, *Né qui, né ora: peripezie mediali della performance contemporanea*, Ledizioni, Milano, 2022, p. 22.

59 The term "digital human" does not have a single definitive originator, but the concept is closely related to the broader idea of augmenting human capabilities through technology, which dates back to the early 1960s. Specifically, Douglas Engelbart's seminal work on "augmenting human intellect" in 1962 laid the foundational ideas for enhancing human abilities with digital tools and technologies (Frontiers Research Topic, 2023). Engelbart's vision has evolved into contemporary understandings of digital humans, encompassing various aspects such as augmented reality, artificial intelligence, and human-computer interaction.

60 S. Hagemann, I. Pluta (eds), *Quels rôles pour le spectateur à l'ère numérique?*, Épistémé, 2023, pp. 61–62.

61 John P., *Digital Humans Explained: What They Are, and How We'll Interact with Them in the Web 3 Age. An In-Depth Guide of Digital Humans*, 14th January 2022: https://medium.com/@oortech/digitalhumans-explained-what-they-are-and-how-well-interact-with-themin-the-web3-age-d2df72cc0425, [accessed 20 March 2024].

62 P. Phelan, *Unmarked: The Politics of Performance*, Routledge, London 1993.

63 P. Machado, J. Romero, A. Cardoso, *Artificial Intelligence and the Arts: Toward Computational Creativity*, Springer, Cham, 2023.

64 G. Kirkpatrick, *Aesthetic Theory and the Video Game*, Manchester University Press, Manchester, 2010.

2 Performative Migrations

Between Dissolution and Metamorphosis

This chapter addresses the evolution of live theatrical performance, now incorporating technical elements such as pre-recorded visuals and motion capture, thereby challenging traditional concepts of "live" theater. This redefinition emphasizes the audience's experience and the real-time adaptability of performances by design teams. Unlike conventional theater, digital immersive content faces constraints that are difficult to navigate. Aesthetic approaches in this context are considered anthropological, transcending cultures and rooted in human evolutionary history. Virtual theater blends real emotions with unreal scenarios, transforming perceptions and sensations. This interdisciplinary analysis, supported by cultural anthropology and pragmatist aesthetics, highlights the active and productive nature of theatrical aesthetics, integrating technology into a dynamic performative space that parallels physical stages.

2.1 Redefining Performance in Virtual Reality

The dimension of a live theatrical performance has, for many years, often been highly technical, incorporating pre-recorded visual and audio experiences designed for repetition or interactive dematerialized scenography. We pose the question: what is live theater, and how does it connect to live-streamed digital experiences, particularly those utilizing motion capture? Although the concept of "live" in this context refers to the audience's experience and their relationship with the performer, there is another aspect of "live" that influences the process of creating and staging a piece of digital theater. The ability to influence and determine outcomes through "live" changes to ideas on stage by the theatrical design team has become a significant part of the production's iteration. This is part of the traditional theater-making process that does not easily adapt to the constraints of digital immersive content creation.[1] The logical theatrical approach chosen for this section of analysis is undoubtedly of an aesthetic nature, considering the latter term as an anthropological fact, transcultural, and the product of the evolutionary history of the human species. This assertion, seemingly entirely legitimate and logical, poses a series of important theoretical problems. Contemporary debate has internalized the

DOI: 10.4324/9781032676937-3

idea that the work of art is not the privileged object of the aesthetic relationship, which can be activated by any type of object and event. Considering that the general and traditional identification of aesthetic and artistic has been overcome, we can still affirm that the artistic and aesthetic spheres are, from a strictly anthropological point of view, connected, and that only in Western modernity does this connection begin to be described as a relationship of identity.[2] The aesthetics of theater, or as we prefer to say, the anthropology of aesthetics concerning the artistic phenomenon of new technologies applied to performing arts, refers to a material or immaterial place of an experience, yet authentic and complete, of truth linked to perception and sensation. Aesthetics has always oscillated between the existence and non-existence of a true epistemic object, between the theory of art and understanding through art, between aesthetics as science and aesthetics as critical philosophy. We prefer to use the term that associates anthropology with aesthetics, referring to a peculiar ecological relationship between humans and the world founded on specific cultural artistic dispositions and realizations.[3]

We will be supported by the concepts of cultural anthropology and pragmatist aesthetics, which are not notions tied to a simple mode of feeling but are primarily specific qualities of action, an anthropology of performative aesthetics that benefits from concepts capable of highlighting the active and productive nature of theatrical aesthetics as behavior. By cutting the notion of conduct, we refer to a sequence of actions significantly coordinated and situated in a specific context, characterized by a unified global direction so as to emphasize the inherently socially significant nature of aesthetic interaction between humans and the world, between the performer and the apparatus.[4]

The mediatized culture,[5] through the democratization of technologies, extended reality, and the metaverse, has demanded the artistic sector to drastically rethink the modes of audience engagement/disengagement, especially during the pandemic, when artists and cultural organizations interacted with their audiences through interconnected and hybrid platforms. Beyond the radical metamorphosis of the author-performers, last but not least, the 21st-century spectator is often disoriented while ‘crossing’ the boundaries of their role, body, or space.

Regarding the essence of theater, we have seen that the concept of theatricality and some traces of ritual seemingly remain. As in every community,[6] ritual behaviors primarily aim to be observed, to appear “spectacular,” without distinction between those who act and those who observe. Similarly, in virtual theater and immaterial performative scenarios, the experience is shared – often simultaneously, although not in the same place – by multiple users/individuals, eliciting real emotions – arising from the sight of unreal places or the transformation of our appearance into artificial beings. Therefore, virtual theater hosts a paradoxical presence that manifests in its physical absence and materializes in a sort of evocation. Being is in its becoming, and the subjective memory of the event becomes collective and sometimes fixed and recorded in a machine, which can lead to the risk of gradual erasure and removal of one's socio-ethnic memory, leading to the emergence of entities without a defined origin and identity.

What seems to have driven humans, even more so during the confinement period of the Covid-19 pandemic, is their unstoppable desire for contact and relationship; what Lacan defines as the "myth of the lamella". Leaving aside the obvious sexual connotation of the lamella that corresponds to the libido, let us focus on how this corresponds to the pure instinct of life, that is, of immortal life, of unconstrained life, of life that, for its part, does not need any organ, of simplified and indestructible life.

> [...] Let us envision that each time the membranes rupture, a phantom emerges from the same breach, representing a form of life infinitely more primordial. The act of breaking the egg gives rise to both *Homo* and *Hommelette* [...] Consider it as an expansive crêpe that moves akin to an amoeba, ultra-flat so as to glide beneath doors, omniscient as it is driven by the pure instinct of life, and immortal due to its ability to reproduce through fission.[7]

Although we are immersed in the performance, sensually enveloped by the environment, we are actually outside the process and enjoy it solipsistically, often relying on and completely delegating, often unconsciously, our drive for contact and relationship to technology. The sense of ubiquity is disorienting. We know how to navigate and operate in virtual space, but we do not know exactly what it is, what it is made of, and the boundaries of its existence. This excess of proximity prevents an objective detached judgment because the most obvious, ubiquitous, and important realities are often the hardest to see and discuss.

Let us now revert to the paradoxical anthropological dimension. The paradox concerns the delocalization and dematerialization of a collective ritual, that of "going to the theater" but also that of creating together, of theatrical reciprocity. Especially since the recent pandemic period, we are witnessing a true migration of bodies and the dematerialization and delocalization of artistic creative processes. Dramas and choreographies are not only accessible remotely, but they are also created among individuals who have never physically met. Once again, we are faced with a new way of understanding the concept of embodiment, crucial in contemporary performance studies. It no longer concerns only the human body in its materiality, but it is a perspective that views cognition as a distributed system, in which culture, mind, and environment are situated in a context of mutual coevolution. "After centuries of being fully clothed and enclosed in a uniform visual space, the electric age introduces us to a world in which we live, breathe, and listen with our entire epidermis".[8] This thought, now almost half a century old, remains relevant when we consider how the art of theater—the art of seeing, of catharsis, of feeling—has been contaminated and enhanced with the introduction of new technologies.

Turning now to technological theater, particularly one in which the viewer is bound to multimedia display devices. In virtual theater, the limit of the image is overcome, and imitation becomes creation if one understands the concept of "virtual" – a word derived from the medieval Latin *virtualis*, in turn, derived from *virtus*, strength, power – as something not opposed to reality and that actualizes without being transmitted in concrete or formal materialization.[9] It creates an image that is not an imitation of someone or something but becomes the instrument that makes the invisible visible. Thus, overcoming the idea that images are a metaphysical reality that does not correspond to any experience, or that the material indeterminacy associated with the virtual corresponds to simplicity in form, the technological device must no longer be considered merely as a tool or aid but as a co-protagonist and creator of the performance. We witness the overcoming of the screen frame's boundary toward immersion in a reality where the retina and the screen merge.[10]

As an experimental artist of these processes, we can cite Gilles Jobin with his company based in Geneva, Switzerland.[11] The live digital performance *Cosmogony – Live Remote Digital Performance*[12] (2021) recalls an artistic experience initiated by the company in 2020 with the project *Virtual Crossings*[13] (2020). *Cosmogony* is a choreography by the Swiss artist Gilles Jobin, a 30-minute show danced by 3 dancers whose movements and bodies are processed into digital bits at the #Studios44MocapLab[14] in Geneva. These bits are then transmitted in real time through cyberspace and appear as avatars on the screen of the audience present in the theater, thousands of kilometers away from the production studio. This performative space is not just a concept but a place of existence and energies. Unlike the *digital humans* seen in previously analyzed performances, the avatars in this performance are not autonomous 'doubles' but are entirely dependent on human gesture and movement. They are an extension live in the metaverse.

> where they function not as "replicas" of real people, but as AI-backed non-playing characters (NPCs), who, unlike their video game counterparts, can react intelligently to your input — your mood as conveyed by the tone of your voice, your facial expression and your body language, for example — and respond to those cues with expressions of their own, lending digital humans a more nuanced emotional and cognitive understanding, as it were.[15]

Cosmogony is a particularly suitable title for a dance performance in the metaverse for several reasons that resonate with themes of creation, transformation, and interconnection. First, mythological cosmogony deals with the birth and development of the universe, concepts that can be easily compared to the creation of digital worlds in the metaverse. In the metaverse, new virtual environments are continuously created and modified, mirroring the dynamism and creativity inherent in ancient cosmogonic narratives.[16] Dance, as an art form, can interpret this dynamic creation through movements that symbolize the formation of new digital realities, staging a parallel between the mythological creation of the universe and the virtual construction of the metaverse.

Second, the concept of cosmogony implies an element of transformation, where primordial chaos evolves into an ordered cosmos. This transformation is a powerful theme that can be explored through dance, especially in a virtual environment that allows infinite creative possibilities. In the metaverse, dancers can leverage technology to transform their bodies and environments in ways that would be impossible in the physical world, using avatars and digital settings to represent metamorphosis and continuous evolution. The body duplicates itself and becomes an instrument of what anthropologist and theorist André Lepecki defines as "politics of the immaterial",[17] describing how bodily performances can challenge materiality and physicality by emphasizing immaterial aspects such as energy, movement, and presence. This approach shifts the focus from the physical form of the body to its expressive and symbolic capabilities.

The term "avatar" originates from the Sanskrit *avatāra* employed in Brahmanism and Hinduism to denote the descent of a deity to earth. Traditionally, this concept is associated with the ten incarnations of the god Vishnu, symbolizing various forms through which the deity intervenes in the terrestrial realm.[18] Beyond this religious connotation, the term has evolved to encompass broader meanings such as reincarnation, return, or transformation.

In the digital domain, the avatar represents the alter ego of the user, serving as a virtual representation of a physical entity, which may be a real person or a complex system. This concept of the avatar is situated within a cultural tradition that includes the notions of "daimon", "shadow", "double", and "doppelgänger" – terms that, in various contexts, express the idea of a copy or a parallel manifestation of an individual. In psychoanalytic literature, for example, the "double" is frequently associated with themes of split identity or self-reflection.

In contemporary metaverses, the avatar assumes a central role, not only as a digital expression of the user but also as a crucial element of interaction and personalization. The customization of avatars in these virtual spaces is one of the principal functionalities, fostering a dynamic and expanding market focused on accessories, aesthetic traits, and features that can be acquired or modified to enhance or alter the avatar's appearance.[19]The evolution of the concept of the avatar, from its original significance in Indian religions to its application in contemporary digital contexts, reflects a wide spectrum of cultural and technological meanings that continue to expand and influence the modes of social and personal interaction in new media.

This type of performance can offer the audience a fascinating insight into how ideas of order and chaos and creation and destruction can be represented through an immersive and interactive experience.

> [...] peoples of the world, they each have their own COSMOGONY. They have a story about the creation of the universe, but [it] is also the study of the universe, like the real one, the physical one. So, it's something a little bit in between. Also, when you choose the title, you need to have [one] that is kind of generic enough, not too precise, you know, I [am] playing [with] technology and really prototyping. You [are] never sure what you're gonna get.[20]

The dancers are "captured" in Geneva and transmitted instantaneously through the network to reappear in real-time as avatars on the large screen. This represents a new generation of "telematic" performance for an audience physically present in the theater yet situated hundreds of kilometers away from the actual scene. Consequently, the question naturally arises: why create a live performance when no audience interaction is required?[21] The Swiss artist addresses this query as follows:

> Choreography is a language that fits new technologies very well, because it is concrete and abstract, it is not narrative but it carries meaning [...] Technology offers new possibilities but also has limits. But a theatre is a limited space in itself: the audience is frontal, and the space available for the performers, the stage, is contained. [...] These limitations force us to be hypercreative in order to create endless stories in a limited space. [...] Errors and glitches are not acceptable in a recording. In a real time performances glitches are only an instant, a momentum. In a way, in a live performance, the audience is expecting inaccuracies, that's what makes us feel alive. Also, something that is often overlooked in live performances is, they get better after each performance because we can rework or edit the piece.[22]

The dancers and their avatars are "particles" in a quantum state, and their movements are teleported instantaneously to any location, akin to parts of an infinite entanglement.[23] Technology does not manage everything; they are the puppeteers of their own bodies, animating their avatars in real time and thus composing the cosmogony of a suspended world, allowing the audience to experience a unique sensation in an expanded space. The synthetic creatures dancing in the metaverse lack identity and can also be considered interchangeable as abstract beings without context and ideal figures that have no history and leave no trace, but rather dematerialize. We can, therefore, speak of metamorphosis, of that place and manner that the virtual renders possible by

Figure 2.1 Cosmogony—Cie Gilles Jobin Dancers: Susana Panadés Diaz, Rudi van der Merwe, József Trefeli. Photo by Cie Gilles Jobin.

reconciling identity and change; where everything perishes, transforms, and renews its appearance.[24] The body is dismembered as a persistent sensation but prolonged as a visual reconstruction, inhabiting another world – an immaterial universe (Figure 2.1).

> We use an optical system. We have about 40 cameras, from Qualisys. The dancers, they have markers… they [are] just reflecting the infrared actually. And the cameras catch it from all different angles and create a skeleton. On top of the skeleton, you can base another avatar. We connect this directly, live, into Unity. When I move under the mocap system, then my avatar is being animated into Unity. In real time. And then from there I can have a real time output. It's basically like online gaming: you have your own PlayStation at home. You generate the game locally and then you connect to a server, and then that's how you grab the position of the other players in the game and you share your own position.[25]

The theater without flesh-and-blood actors has existed for centuries, from the Baroque period to the Bauhaus and even more recent object theater; the fundamental significance of the actor's presence in theater has therefore not been undermined by new technologies and these new scenic forms. The split between the execution and perception of the gesture, between the sensation of the instantaneous "now" and the projection of the self into a dematerialized environment, largely depends on the audience's awareness. A first logical observation concerns movement: in Virtual Reality (VR) technology, it is drastically separated from the somatic awareness of the dancers. While dancing, performers usually focus their attention on their bodily sensations to execute and complete movements effectively. Staging dancers through a VR system thus presents a challenge for their perception, as the virtual environment 'converts' and diverges from the canonical choreographic experience.[26]

> Our experience showed us that motion capture must be investigated in movement. Cleaning mocap data is very expensive, experimenting with time inside the mocap space can save a lot of expense in post-production. Every mocap system has its own shortcomings. Optical, accelerometers suits, point cloud mocap: creators have to find the most suitable mocap system for their project while movers need time to understand their specificity and there are many solutions that can be found in movement. As I often say, we must find artistic solution to solve technical problems.[27]

As performers move in a virtual environment, there is an interruption and discontinuity between what the dancers see and what they feel in/on their bodies. Initially, it might seem that VR technology leads the performer toward an artificial dualism, a Cartesian mental state, where visions, actions, and thoughts are disconnected from the fleshly body. The real performers stand alongside virtual avatars, in a configuration that demands various types of writing: the scenic writing that unfolds on the stage, the writing tied to the interpretation of actors and avatars, as well as the computer-based writing created with the aid of different software, enabling the connection of these diverse components. Indeed, dramaturgy finds a new dimension in the image and in what is termed digital writing, linked to binary code and hypermedia devices.[28]

> This allows for the generation of new textual and hypertextual qualities. Technological devices in theater thus pose a fundamental challenge for the actor: they are confronted with the image or the digital object as if they were their stage partners, experiencing a 'technological augmentation' on a stage populated of robots, sensors, projected images, and they face a new scenic situation where space and time are multiplied, where they are filmed, and where they must position themselves simultaneously in relation to the camera and the audience.[29]

The actor/performer, who inherently embodies mimesis and creates 'appearance', delivers their Self and their embodied intelligence to a digital spirit, to their phantasmagoric double, transcending spatial boundaries and even defying the laws of gravity. The audience is led on a journey into a suspended reality, rich in poetic and aesthetic charge, and full of a fantastic dimension. But can the same be said for the performer's experience? The performer knows they are being watched but does not feel it; the reciprocity and the relationship dear to the live performance stage are missing. Technology has enabled the construction of a bridge between the physical and digital worlds, giving rise to the dimension known as *phygital*,[30] where performative environments can be occupied and inhabited not only physically but also, and especially, psychologically, by both the performer and the audience.[31]

The sources of the virtual setting of *Cosmogony* were the numerous films shot during the pandemic, with footage of empty streets and entire cities, captured by drones. In this desolate scenario, fantastical means and entities are inserted, digital metamorphoses of the dancers, and each spectator has a unique view in relation to their position. At the beginning, middle, and end of the performance, the dancers in flesh and blood are shown, wearing their sensor-covered suits. However, this alone cannot save that primordial quest for notoriety so dear to the performer's nature, the desire to bring forth their artistic subjectivity.

> In the case of the virtual body, the event is unrepeatable particular that is constituted as a system *at least* by the interaction between a human body (thus of a mind-body plexus) equipped with technological prostheses, and an electronic processor implemented by an algorithm [...]. Now, does such an individual exclusively occupy a place? If so, what place? Certain parts of my technological prostheses, certain superficial sensibilities of my body, a certain part of my brain, a computer memory? It is in any case a body that admits other bodies in its place, admit to being traversed [...].[32]

A suitable starting point for analyzing these unique theatrical objects is Clifford Geertz's classic work in modern anthropology. Geertz, in defining the origin of culture, speaks of symbolic objects:

> thinking does not consist of 'events in the head' (...) but in the traffic of (...) meaningful symbols - mostly words, but also gestures, drawings, musical sounds, mechanical devices like clocks or natural objects like jewels - anything that is detached from its simple reality and used to confer meaning on experience.[33]

We can associate the virtual reality helmet with that particular concept of cultural technology coined by Geertz, which for centuries has been magic, seen as a set of practices capable of shaping objects with such strong symbolic power that they influence the physical world itself. This is valid both from an ethnological perspective, where rituals are always activated through particular objects, and from a narratological perspective. The helmet worn by the dancer or the spectator acts as a narrative activator and is capable of triggering in both the performer and the observer the emotion of witnessing the initiation of a magical plot.[34]

The performative domain, in all its variants, is shown as a practice in which ideation and action are inseparable, behavior and reflexivity develop together through that notion of "practical idea" developed by Marcel Mauss regarding magic.[35]

> An idea that is also an effect of the situation, within the body in action, within a certain narrowly and collectively defined field of experience, capable of exerting its influence beyond that narrow scope, opening it up to the broader reality as a whole. This applies to ritual, magic, and theatricality, although differently for each experience.[36]

Emerging trends in immersive performances have been theorized to consider the spectator's body and its perceptive faculties as the primary locus for creating artistic meaning.

> VR is much closer to performing arts because you cannot force people to look in a certain direction. Depending on where they stay – more in the front, middle, back – people don't have the same experience. That is something that for us is quite natural to take into account.[37]

The focus is no longer solely on the performer. The centrality of the audience's body has led to increasingly widespread ontological claims in favor of immersive theater, promising that the 'haptically' embodied spectator can more fully experience various phenomena and identities.[38]

The work of the Swiss artist is currently on an international virtual tour – *#tourwithouttraveling* – and has already been broadcast live from Geneva across three continents. Jobin aimed to preserve the live aspect by ensuring that each performance is unique; the version presented in Singapore in May 2021 was followed the next year by the one presented at the *Sundance New Frontier*. While maintaining the same accompanying music and duration, the choreography and camera angles were adjusted to better define the performers' movements, and the appearance of some avatars was altered, giving them human-like eyes.

> The venue that invites COSMOGONY must provide a good PC, we send them the build of the piece beforehand, they generate the build locally on their PC and they connect the projector or another video output. It requires a stable internet connection to connect to the server to grab the mocap data live from our Geneva studio. So that's why it's very light and, and there is no lagging with an HD quality screening. We developed a tool for this technology that allows us to invite any motion capture system into a 3D space such as Unity. So you can be from South Africa in, uh, a Rococo suit dancing with (dancer) Susana here in Geneva, and somebody somewhere else dancing together but distant from thousands of miles away. This is feasible today and opens many creative possibilities.[39]

Cosmogony represents one of the numerous recent artistic cases where live performance, the art of actualization, and the co-presence of material forms summon virtual and 'mixed' presences, attempting to control them like puppets or extraordinary creatures that can be manipulated and traversed. We are faced with one of the many cultural paradoxes of the new technological artistic trend: the ability to be dynamic in complete stillness and to travel and be everywhere without moving.

In agreement with what Auslander asserts,[40] when analyzing the inevitable fusion and overlap between live and mediatized performance, Cosmogony demonstrates how the characteristics of immediacy and intimacy, once exclusive to live performance, can also be associated with other media, profoundly altering the perception and value attributed to physical presence. Auslander emphasizes the need to overcome a binary vision that opposes the two modes of experiencing performance, inviting a broader reflection that transcends simple ideological and cultural dichotomies.

This approach fosters a balanced and critical interpretation of the type of performance chosen in this analysis, which clearly illustrates the shift to an art that, in addition to addressing the invisible, becomes a total experience, understood not as a void absence but as the condition of every presence, as a force.[41]

2.2 Living (in) Another's Body

Theater has always been a place where one can assume new identities, hide your own appearance, live the lives of 'others,' and indulge in dreams. This illusory play has now been taken to the extreme by human knowledge, occurring with such a level of participation and immersion by the audience that a significant number of individuals no longer wish to live in the real world. Instead, they spend hours daily in that world where everyone can be a great actor. The intensity of certain technological artistic experiences is directly proportional to the degree of authenticity of some performative experiences and their representation of reality. As may already be clear, this study will not follow a technophobic line that seeks to demonize Extended Reality (XR) tools—technologies that include all immersive technologies that extend our real world and combine it with virtual elements—but will also not paint an idyllic picture of the effects of current artistic and spectatorial trends.

We live in a performative era that, beyond constant metamorphosis, shows the fading of corporeality on stage or, perhaps more generally, of the human being. The concept of fading is often associated with ideas of transition, transformation, and loss of definition or clarity. In philosophy, fading symbolizes a movement toward less definitiveness, highlighting processes of change, weakening, or transformation that can be applied to multiple dimensions of reality and human thought.[42] A pertinent example is Simon Senn's second lecture performance, *Be Arielle F.*[43] (2020), where the Geneva-based performer tells the audience how he purchased the naked body of a digital avatar, modeled on the body of a real English woman, from the website 3dscanstore.com.

> It all started when I met Arielle
> Well, not the real
> It all started when I met
> Her digital double, her virtual replica.
> [...]I was testing an immersive
> motion tracking system in virtual reality (VR).

> For those tests I needed a virtual body that shared... More or less the same body proportions as my own. [...] I checked their catalogue of male bodies but couldn't find one in my size.

> then checked the female bodies and chose Arielle's digital
> [...]I needed to find Arielle and talk about what I'd felt in her.[44]

The file contained a static, photorealistic three-dimensional representation of the naked body of a young woman. Senn used a free online tool to give it digital bones, allowing him to make it move. "I mean, I recognized my gestures even though it wasn't my body".[45] He bought virtual reality equipment, including sensors intended for video games, placed them on his body, donned the virtual reality headset, and 'became' a young woman. The performer directly and succinctly narrates this experience, mirroring the Proteus effect that virtual environments, such as online games and web-based chat rooms, allow users to dramatically and easily alter their digital self-representations. When we change our representations of the self, our self-representations, in turn, change our behaviors. This becomes evident in the second part of the performance when the real Arielle asks *him/her* to dance to the music of The Pussycat Dolls (the song *Don't Cha*).

The Proteus effect[46] makes us reflect on the fact that the visual characteristics and traits of an avatar are associated with specific stereotypes and behavioral expectations. When an individual believes that others will expect certain behaviors because of the appearance of their avatar, that individual will strive to adopt the expected behavior with their avatar. Senn brings the results of this effect to the stage, winking in his grotesque and sexualized dance, adopting movements and expressions that he believes belong to the female gender.[47]

Simon recounts his experience to the audience of "being in the body" of another person, describing how he felt moving through his transgender clone,[48] with his face and a female body, bringing to the stage issues related to the contemporary theme of "fluid" gender identity. This concept recognizes that gender is not a fixed and immutable category but rather a dimension of

Figure 2.2 Simon Senn and "Arielle"—Be Arielle F, Simon Senn, Photo ©Mathilda Olmi.

human identity that can vary over time and in different contexts.[49] This notion challenges the traditional binary view of gender, which rigidly distinguishes between male and female, and proposes that people can experience a range of gender identities that do not necessarily conform to these categories.

After direct dialogue with the audience, he then calls – via a FaceTime call – the woman to whom that body belongs. Together with 'Arielle', they reflect on the ethical, existential, legal, and psychoanalytic questions related to this experience. Initially acting as a mere puppeteer, then animating his virtual puppet by giving it the "breath of life" through technology, the performer co-constructs his fictional identity (Figure 2.2).[50].

This dynamic is completely altered when one chooses to live an experience through an avatar, whose appearance and image is effectively an eidolon that dulls minds, assuming an appearance perceived as real and magical by those who are its simulacrum. "Our current common 'environment', in which we constantly seek adaptation and rebalancing, now has the traits of the media imagination and the related regime of desire typical of the era".[51]

We are witnessing a process of mediated identification, explored by Turkle[52] in the mid-1990s,[53] which refers to the way individuals interact with and identify through technology, particularly through computers and the Internet, which serve as mediation tools. Turkle argues that computers and virtual environments create spaces where people can explore different aspects of their identities in ways that are not possible in the physical world. She suggests that these mediated environments offer opportunities to experiment with identity, allowing individuals to "try on" different personalities, roles, and characteristics in a reversible and safe manner. In virtual spaces such as online communities, chat rooms, and multiplayer games, users can present themselves differently from their real-life personas without compromising their social masks. This ability to create and manipulate digital avatars allows people to explore aspects of their identities that may be suppressed or unexplored in their offline lives. These virtual experiences can lead to a deeper understanding of oneself and others, as individuals reflect on their virtual interactions and how these relate to their concrete social relationships. Turkle emphasizes that these virtual interactions are not merely fantasies of escapism but can have profound effects on the psychological and social development of an individual.

Another key point in understanding the function of simulating a new performative identity is linked to the notion of hyperreality, a state sometimes described as a cause of the "decline of the humanist era".[54] It is established in a society weakened in its imaginative capacity, favoring an excess of information and codes, where a weakened, if not corrupted, rationality emerges, reproducing ineffective processes of copies at an uncontrollable speed. Hyperreality is established when there is a subversion of signs intended to represent reality in an ultra-symbolic form but have lost all contact with it.[55] If no reality exists, one might spontaneously think, then the hyperreal universe might refer to the faculty of the imaginary, to fantasy, which through the dialectic of creation and dissolution invents its representations. Yet, while hyperreality detaches from reality, simulating it, it simultaneously strongly avoids the realm of the imaginary, which in such a hyper-informational society does not find the right space to emerge.

In theater, hyperreality can be achieved through various techniques such as advanced scenic design, multimedia elements, virtual reality, augmented reality, and interactive storytelling. These elements create an intensified sense of immersion for the audience, blurring the boundaries between the physical world and the imaginary world represented on stage. Hyperrealistic theatrical experiences aim to evoke intense emotional responses and deep engagement from both the performer and/or the spectator, making them feel as if they are directly experiencing the events unfolding on stage. This can lead to stimulating reflections on the nature of reality, perception, and the role of technology in shaping our understanding of the world and our subjectivity.

The concept of 'hyperreality in theater'[56] raises ethical questions regarding manipulation and deception, as well as concerns about the loss of authenticity and genuine human connection in an increasingly mediated world.[57] Despite these challenges, hyper-realistic theater continues to push the boundaries of artistic expression and audience engagement, offering unique and immersive experiences that challenge our perceptions of reality. However, it is clear that the described performative experience is not at all linked to a state of depersonalization, considering the emotional storm that invests the artist during his experiment, but rather to a phenomenon to which we are all now accustomed, derealization, in which time can appear suspended or incredibly fast and the surrounding world can appear. The technological potential relates to the artist's expressive capacity and, equally, the expansion and development of the imaginative capacity of the involved audience.

The philosopher Nelson Goodman contended that emotions are not merely passive experiences but active cognitive processes essential for interpreting and engaging with the world. Emotions, he posited, function as modes of understanding that shape our perceptions and interpretations of various situations and experiences. As Goodman stated, "Emotions function cognitively in organizing and interpreting experience".[58]

Goodman's analysis extends significantly into the realm of fiction, where he emphasized the crucial role of fictional works in constructing symbolic worlds. Fiction, according to Goodman, is not simply an escape from reality; it is integral to our comprehension and engagement with the world. Through fictional narratives, we explore possibilities, test hypotheses, and broaden our understanding of diverse perspectives and experiences. Goodman argued, "Fiction provides a way of worldmaking that is no less significant than the scientific approach".[59]This engagement with fiction facilitates the construction and navigation of complex symbolic worlds that inform our real-world interactions and perceptions.

Goodman's insights are particularly relevant in the context of theater in virtual reality (VR). VR theater represents a convergence of technological innovation and artistic expression, creating immersive environments that enable participants to engage with fictional narratives in highly interactive and sensory-rich ways. These VR experiences utilize the cognitive functioning of emotions by immersing users in vivid, emotionally charged environments, thereby enhancing their understanding and engagement with the narrative.

The concept of fiction as a constructor of symbolic worlds is vividly illustrated in VR theater. Virtual spaces allow for the creation of entirely new worlds, where the boundaries between reality and fiction blur. Participants navigate these worlds, interacting with characters and environments in ways that evoke strong emotional responses and facilitate deep cognitive engagement. This immersive experience underscores fiction's role in expanding cognitive horizons and enhancing our capacity to interpret and understand complex symbolic systems. As Goodman noted, "The worlds of fiction and reality intermingle, each enriching the other".[60]

In essence, the application of Goodman's theories to VR theater highlights the powerful interplay between emotion, cognition, and fiction. It demonstrates how virtual environments can serve as platforms for exploring and constructing symbolic worlds, offering participants a rich tapestry of experiences that inform and enrich their real-world understanding. By engaging with these virtual narratives, individuals not only experience heightened emotional and cognitive responses but also contribute to the ongoing construction of their symbolic worlds, reflecting fiction's profound impact on shaping human cognition and perception.

In an era marked by artistic (and not only) paradoxes, where the medium is not just the message, quoting the esteemed McLuhan, but also defines our situation and being.[61] The essence of theater remains in the concept of theatricality and traces of ritual. In every community, ritual behaviors are primarily meant to be observed, to appear "spectacular", without distinction between actors and observers. In virtual theater and immaterial performative scenarios, the experience is shared by multiple users/individuals, often simultaneously, albeit not in the same physical space, resulting in genuine emotions, albeit derived from seeing unreal places or transforming our appearances into artificial beings. Thus, virtual theater hosts a paradoxical presence that manifests in its physical absence and materializes in a form of evocation. Being is in its becoming, and the subjective memory of the event becomes collective and, sometimes, fixed and recorded in a machine. This may lead to the risk of a gradual erasure and removal of one's socio-ethnic memory, resulting in the emergence of entities without defined origin and identity.

This human and, therefore, sometimes artistic submission, is not a new phenomenon, as noted in the already mentioned volume where Aristotle described slaves as "living tools" of action: "Every piece of property is a tool for life, and property is an assemblage of tools: even the slave is an animate piece of property".[62] These considerations clarify the nature and essential quality of a slave. A being who by nature does not belong to themselves but to another, although they are human, is by nature a slave: and belongs to another who, although they are human, is an object of property: and an object of property is a tool directed to action and separate.[63]

Despite the knowledge of such sophisticated technological tools, awareness is sometimes lacking because in the attempt to escape anonymity, one ends up in a kind of homogenization where the peculiarities characterizing the individual are stylized. There is a continuous transition from one generation of reality to another, through a movement of derealization articulated in two main phases: a phase of concrete reality, concerning the field of philosophical, scientific, or artistic representations, and a phase of generally unnoticed substitution, where the real of the previous generation is replaced by the real of the new generation.[64]

Normally, in fact, the only conditions through which an environment and an ecstatic experience are perceived as more factual than reality are usually dream or hallucinogenic/psychotic experiences, and theater, linked to the condition of the here and now, leads to a purely conceptual there and then. With the integration of the most advanced technologies in the performative field, the claim of "transformation" or physical immersion "in" a distant elsewhere is taken to the extreme, although it is indeed related to the participant's attitude. The more the system is susceptible to malfunctions—think of a VR headset malfunction or a bug in the AI system—the greater the risk of a collapse of the experience, resulting in the agent's frustration. Behind the promise of a show and a real experience—and at the same time impossible in everyday life—lies the risk of finding a disobedient or refractory audience, or in the case of younger people, completely dependent on and accustomed to such dynamics to the point of no longer finding any difference between the concrete and virtual act. This is not an entirely empirical problem but rather an open one, without a univocal resolution that embraces our current mental resources and logical constructs, but which opens up to a certainly philosophical debate due to the category of the discussed problem, involving this rapid and radical digital transformation of which we are protagonists, not entirely aware.

Steve Dixon describes this widespread technological trend in the field of Performing Arts through the philosophical lens of Cyber-Existentialism.[65] The two fields are associated through procedural analogies and find their synthesis in artistic examples like those previously proposed. Existentialism highlights personal freedom and our ability to reinvent and recreate ourselves continuously, while cybernetics seeks to create systems and environments that have the freedom to be flexible, adaptive, or autopoietic—self-generating or self-regulating. Autopoiesis is based on both unity and circularity, as in the human nervous system, which triggers and stimulates the spontaneous production of actions or substances to correct and regulate the system or organism: it is the circularity of its organization that makes a living system a unit of interactions, and it is this circularity that must be maintained to remain a living system and preserve its identity through different interactions. These principles are also central in the field of interactive arts and performing arts that communicate through new technologies.

The concept of live presence has always been considered the specific characteristic of theatrical art, unlike painting, sculpture, or cinema, as its essence is created night after night in front of an audience. However, the ideology of live presence might be based on a misconception, as live presence is not a phenomenon of physical reality but a mental phenomenon, a matter of awareness and energies. "Being present, presence, is a timeless process of becoming aware: in other words, a process simultaneously inside and outside the flow of time".[66]

On stage, Simon Senn demonstrates to the audience the intensification of the performer's presence, showcasing his exceptionalism as it manifests and occurs. We do not witness a representation but a presentation, an exposition, a presence that is also spatiality and an approach to sharing the scenic space[67] with the performer and his virtual double. The empathy established through the theatrical performer-audience relationship, which refers to that form of immediate, physical, and emotional identification—almost an inhabiting the body of the person in front—is transformed into an embodied consciousness on the part of the spectator who witnesses the detachment of the performer's identity and body in favor of the creation of an image of a new self.

This brings us to that *performative turn*,[68] which in contemporary culture and society is often the result of the pervasive mediatization of culture and society. The media have become a substantial part of reality itself, more than merely representing reality through a mediating function.

> In other words, our mediatized culture and society have turned into a hyperreality of simulations and simulacra, which means that the signs have become more real than the objects to which they refer or, to put it differently, that reality has been replaced by its representations.[69]

2.3 Digital Puppeteering: The Transformation of the Human Body through Emerging Technologies

Along with the performative forces, intermedial dramaturgies crisscross simultaneously the actual and the virtual[70] where digital alterability meets performative proximity. The confrontation between mediatic strategies and physical presentation results in the fact that not only the body is performed, but also the medium performs itself.[71] This is the radical feature of intermedial theater.[72] Upon analyzing certain contemporary artistic practices that incorporate new technologies, numerous aspects suggest the emergence of a relational aesthetic paradigm. This paradigm emphasizes encounters and intersubjectivity, which manifest themselves in new zones of communication and instant collective rituals.[73]

> For the participating spectator involved in the realization of the artwork, it is less critical to possess expertise—defined as having the requisite skills to comprehend and act—than to be open to experimentation, which entails being receptive to potential actions. Within the context of experiments, the forms of action and discourse can be continuously reformulated.[74]

Adapting to the new experiential and real conditions imposed by realities such as cyberspace and virtual reality does not necessarily lead to a homogenization of actions performed by performers and prosumers. Instead, it fosters a creative adaptation of gestures increasingly oriented toward relationality. Undoubtedly, the axiom most profoundly altered, alongside that of time, is that of space. This alteration now yields intriguing outcomes at the intersection of the online and offline dimensions—the so-called *onlife*[75] dimension—and between the physical and virtual realms, referred to as *phygital*.[76] The new performative spatiality thus represents two articulations of an experience and a possibility, and it is configured as an anthropological place of identity, relationship, and history.

In recent years, the discussion on the concept of illusion within the contemporary mediascape has intensified, particularly regarding its relevance in digital environments such as virtual, augmented, and mixed reality technologies. These immersive environments have the ability to evoke a strong sense of integration into almost-real worlds for users. Consequently, despite traditionally carrying a negative connotation, the term "illusion" is increasingly seen in a positive light as a crucial aspect of the phenomenon of immersion. Thus, it is considered an important objective for creators of hyper-realistic and virtual environments to pursue this effect. In simulated virtual environments, people experience a strong sense of presence (place illusion) and react to what they perceive as if it were real (plausibility illusion).[77] At the same time, they remain perfectly aware that they are not "truly" there and that the events are not "actually" happening. How can this conflict between knowing and perceiving be explained, and can it be considered a new form of aesthetic illusion?[78]

The metaverse is no longer an anonymous space of transit or purely instrumental use but becomes a place of investment and artistic perspective. "The metaverse[79] is a combination of two words: meta, meaning virtual and transcendent, and verse (transformation of the universe), meaning the world and universe".[80] The metaverse is not inherently dystopian. This is a common misconception because the term "metaverse" comes from a dystopian novel, *Snow Crash* by Neal Stephenson (1992). Drama underlies most works of fiction; utopias are rarely the setting for popular stories. However, since the 1970s numerous "proto-metaverses" have emerged that do not focus on domination or profit but on collaboration and creativity. Each decade the realism of these worlds improves, as do their functionality, value, and cultural impact.

The metaverse is often described inaccurately. Devices such as the Meta Quest (formerly Oculus VR) or augmented reality glasses like the famous Google Glass are mentioned as primary tools for accessing the metaverse, but they are not the metaverse itself. Similarly, smartphones are not synonymous with mobile internet. The metaverse is also not represented by games like Roblox, Minecraft, Fortnite, or other similar platforms; these are virtual worlds or platforms that will likely be part of the metaverse, just as Facebook and

Google are part of the internet. The metaverse implies and will produce new behaviors and technologies.[81]

The metaverse, and consequently the performances generated within it, are characterized by:

- Persistence: The metaverse environment is active and exists even when it is not "inhabited";
- Simultaneity: It allows a large number of users to interact simultaneously;
- Interoperability: The ability of two or more computer systems to exchange information with each other and to use that information. In building the metaverse, interoperability is essential to ensure an adequate user experience, such as the ability to buy a digital object in world X and use it in world Y, or the ability to use one's identity across all visitable environments;
- Mixed reality: Through the integration of VR, AR, and the internet, to overlay the physical world or exist in entirely virtual spaces.

The intersection of the metaverse with performing arts has certainly yielded results closer to the playful world of video games and experiences similar to role-playing games concerning the spectator. That said, we can still find interesting experiments and fertile ground in the dance and theater domains, particularly regarding storytelling. Interactive storytelling is indeed at the heart of new digital theater formats, including *online* ones. It requires the ability to conceive and organize narrative content that, in a transdisciplinary manner, can bring together classic dramatic structures with the hypermedia and interactive narration typical of digital multimedia.[82]

Technology generates the "semblance (*Schein*) of an actual presence without, however, actually bringing bodies and objects into manifestation (*Erscheinung*) as currently present. Using certain procedures, they manage to formulate the promise of an actual presence".[83]

The focus of visual dramaturgy—one might say—constructed in this immaterial space is oriented on the subject, interrogating its perceptual mechanisms, listening to the dynamics of sensation, and exploring the relational networks that define and constrain it within its environment and in relation to others it faces. The immaterial empty space becomes a container to be filled, enveloping the spectator; the artwork is no longer contained within the space but is the space itself, understood as an environment, in our case, a virtual one.[84] The notion of space tied to a plastic dimension and representation diminishes, giving way to an experiential notion linked to the perception of one's presence.

In this performative context, there is a shift from a representative aesthetic to a behavioral ethic[85] in which the artist, the spectator, and the artistic object are all implicated in a behavioral context that is creative, as well as physical, emotional, and conceptual. The performance delineates both.

Erving Goffman provides valuable tools for understanding how people construct and interpret their experiences, including virtual reality (VR) environments. Virtual reality, with its immersive and interactive capabilities, can be examined through its concepts to reveal how users navigate and make sense of these digital worlds. In a VR experience, the concept of "framework", a mental structure that people use to organize and understand their experiences, is crucial. In the virtual environment, frameworks help users determine what is happening and how they should behave: "a frame is a lens through which the individual observes social reality and defines what is real".[86]

Virtual reality often requires a process of "keying", which Goffman describes as "the transposition of an activity or event within a different frame"[87] in which a situation is transformed and reinterpreted in a different context. For instance, a VR dance experience like *Le Bal de Paris*[88] (2021) by Blanca Li—an *immersive live show* that blends dance, music, and interactive performances to transport participants to a grand Parisian ballroom—is an immersion into an unreal and timeless universe, simultaneously retro, futuristic, classical, and contemporary, created digitally. This hybrid performance mode allows users to be aware of being in a simulated environment while still experiencing a profound emotional and artistic engagement. This process of reinterpretation is fundamental to maintaining the immersion and effectiveness of the virtual experience:

> By "environment," it is intended to place the user in a total perceptual context, psychologically and physically inhabitable, in which they constitute the unpredictable variable, an indispensable active component of the work. In the absence of the user, the work emits no signals and exists only in a potential state; with the presence of multiple users, the quantity and variability of emitted signals increase, thus enhancing the aesthetic information of the work.[89]

Thanks to the use of augmented reality, optical viewers, and motion sensors, the spectator is invited not only to witness the scenic action but also to be a protagonist by dancing and interacting with the live dancers.[90] The narrative begins with a sumptuous ball organized in Paris by the wealthy banker Richard de la Rivière in honor of his daughter Adèle, who has just returned to Paris after years of traveling around the world. Among the guests, Adèle's father introduces her to James Markus, a young and charming businessman. Adèle finds Pierre among the guests; Adèle and Pierre had been lovers in the past, and their relationship ended painfully. Adèle starts dancing first with James and then with Pierre, rekindling the flames of the past that never really went out (Figure 2.3).

Figure 2.3 Dancers Le bal de Paris. Blanca Lì. Image **Unreal Engine Demo**

In the virtual world, participants become the guests of the party. They transform into hybrid and stylized creatures, with a human body and an animal head. Their costumes, enriched with animated and dynamic motifs, were designed by Chanel's creative team in collaboration with Vincent Chazal. This is a virtual Haute Couture collection, specifically created for the performance. The costumes evolve with the aesthetics of each act, contributing to the overall harmony of the work: black and white at the beginning of the first act, floral and organic in the second, predominantly red and green in the final act. By wearing the viewer, each participant must choose the clothing and animal head of the character they will appear as during the experiment. Upon activation of the virtual reality viewer,[91] the first image they will see is their avatar reflected in a large mirror.[92]

> I wanted to create a show that did not yet exist and that I could not have imagined twenty years ago: this is what excites me.[93]

The concept of the mask, elaborated philosophically by Friedrich Nietzsche, can be closely linked to the spectator's adoption of an avatar, as in Blanca Li's performance. Nietzsche explores the idea of the mask as a tool for revealing and protecting authentic identity, asserting that "every profound spirit needs a mask".[94] This suggests that the mask is not merely a means of deception but rather a necessity to protect and manifest the true essence of the individual. In the context of the analyzed performance, the use of digital avatars allows performers to wear a technological "mask" that expands their expressive and creative capabilities. Avatars enable the spectator-actor not only to explore new dimensions of identity and subjectivity but also to transcend the physical limitations of the human body and the constraints associated with fear of social judgment.[95]

> Attendees are encouraged to don any of the suits and dresses available, regardless of gender, as identities are hidden behind animal masks that cover the avatar's entire head, allowing a sense of anonymity and comfort for those who are shy. Some of the most fun moments happen when participants choose to wander outside the VR constructed bounds a little. One audience member is frequently spotted away from the group, standing in mid-air high above the virtual crowd or seemingly standing on water as a boat takes everyone around a lake.[96]

Through their avatar, the spectator experiences greater freedom of expression and action, hiding behind a digital mask that transforms them into a new actor, different from themselves. This dynamic reflects the idea of "play" and "performance" studied by anthropologists[97] who view theater and play as liminal spaces where individuals can explore and negotiate new identities.

The use of avatars in Blanca Li's theater thus represents an interesting fusion of philosophical and anthropological concepts, demonstrating how technology can amplify and transform the performative experience and self-perception. We can define this type of performance as a manifestation of the aesthetic orientation, as theorized by Martin Seel, and consequently, it also liberates it from the constraints of appropriateness.[98]

It focuses on how individuals engage with and perceive the qualities of experiences. This orientation is not merely about passive observation but involves an active, emotionally intensified engagement with the world. According to Seel, the aesthetic orientation allows individuals to perceive and experience the actuality and internal constitution of their own experiences within a specific conduct of life.[99] In extended reality (XR) performance, this concept takes on a new dimension. XR blends virtual and physical realities, creating immersive environments that heighten the sensory and emotional experiences of participants. This aligns with Seel's notion that aesthetic orientation is characterized by an interest in presenting and experiencing the qualities of experiences in a heightened, affective manner. Participants in XR performances are not just passive viewers but are actively engaged in a shared lifeworld, experiencing a sense of presence and immersion that is both individual and collective.

Seel's aesthetic orientation also involves a reflexive awareness of one's subjectivity within a shared communal experience. In the context of XR performance, this is achieved through the immersive and interactive nature of the technology, which allows participants to explore their perceptions and emotions in a space that feels both personal and communal. The shared yet individualized experiences in XR performance mirror the communal experiences Seel describes, where participants can reflect on their subjectivity while engaging with others in the same virtual environment. Moreover, Seel's concept of "de-realizing" objects to make them represent an entire class is particularly relevant in XR performance. In these immersive environments, objects, actions, and events are staged in ways that transcend their physical reality, becoming symbols or signs that represent broader concepts or experiences. This staging process demonstrates an internal logic, as the immersive technology creates a self-contained context that participants can explore and interpret.

Overall, the aesthetic orientation in XR performance involves an emotionally charged, reflexive engagement with immersive environments that challenge and expand participants' perceptions of reality. This orientation fosters a deeper understanding of one's experiences and subjectivity within a shared communal context, aligning with Seel's theoretical framework.

The hybrid show lasts an hour and a half, with a 35-minute session dedicated to a virtual reality segment that immerses the audience in a fantastical and poetic universe. This is followed by a transition into the real world, where the audience continues to dance with the performers, thus maintaining the immersive and festive spirit throughout the show.

Once again, the aforementioned Goffman provides valuable tools for understanding how people construct and interpret their experiences, including virtual realityenvironments. Virtual reality, with its immersive and interactive capabilities, can be examined through its concepts to reveal how users navigate and make sense of these digital worlds. In a VR experience, the concept of "framework", a mental structure that people use to organize and understand their experiences, is crucial. In the virtual environment, frameworks help users determine what is happening and how they should behave: "a frame is a lens through which the individual observes social reality and defines what is real".[100]

Virtual reality often requires a process of "keying", which Goffman describes as "the transposition of an activity or event within a different frame"[101] in which a situation is transformed and reinterpreted in a different context. For example, a VR dance experience like Le Bal de Paris can be a keying of a live performance, where users are aware of being in a simulated environment but still manage to have a profound emotional and artistic experience. This process of reinterpretation is fundamental to maintaining the immersion and effectiveness of the virtual experience.

Another relevant aspect is the "fabrication" of virtual reality, which involves the intentional creation of a fictitious situation that mimics reality. This process underlies the experience created by Blanca Li, where the environment and interactions are constructed to appear as realistic as possible while being consciously artificial. This phenomenon reflects Goffman's idea that "fabrications can be accepted as reality just as much as real situations, if managed correctly".[102] The negotiation of the audience in a VR context reaches profound levels of interaction and engagement. Users not only accept the framework proposed by virtual reality but actively contribute to maintaining and enriching it. This dynamic interaction reflects the human capacity to create and sustain shared meanings even in highly technological environments. In this context:

> the spectator's contribution to the construction of such a model embodies an act of presence that produces two immediate effects: the creation of a collective work and participation in a social moment of sharing (an intrinsic quality of the art venue), which has transactional virtues (any work is a model of a practicable world).[103]

The spectator, as the constitutive element of the work, participates in its realization and facilitates the symbolic completion of the world being created.

Finally, we come to the last suggestion that this performance brings, which is linked to mythology. The depiction chosen for the avatars in this show—human beings with animal heads—is a recurring theme that spans many cultures and artistic traditions, offering a rich field of investigation from an anthropological and symbolic point of view. These representations embody an intersection between the human and the bestial, often used to explore and communicate complex and profound meanings. In many ancient mythologies, these figures are often gods or sacred entities that represent the fusion between the human and animal worlds. For example, in ancient Egyptian civilizations, deities such as Anubis, with a jackal's head, and Horus, with a falcon's head, embodied death and protection, respectively. These representations not only symbolized the qualities attributed to the associated animals but also served as intermediaries between the earthly world and the afterlife.[104]

From an anthropological perspective, the depiction of figures with animal heads can be seen as a way to express the interconnectedness between humans and nature. In animistic societies, where it is believed that spirits inhabit all living beings, animals are seen as bearers of wisdom and power. The Promethean desire, already observed in the artistic creations described above, resurfaces: the representation of humans with animal characteristics can symbolize the desirability of acquiring the qualities attributed to those animals, such as the strength of the lion, the wisdom of the owl, or the agility of the deer.[105]

Furthermore, it is possible to draw a parallel between the use of avatars and the function of technology as a medium, a bridge between two worlds: one natural and one supernatural. Just as figures with animal heads serve as intermediaries between the human and divine worlds, modern technology acts as a bridge between the physical reality and virtual realms. This mediation allows individuals to explore new dimensions of existence, escaping the limitations imposed by physicality and social constraints.

This interaction with technology can be seen as a modern ritual, in which the user, by wearing a virtual reality headset or manipulating an avatar, participates in a symbolic and experimental transformation. Similar to ancient rites that used masks to evoke spirits and transcend everyday reality, the use of technology creates a liminal space where identities can be flexible and where the concept of self can be explored in new and unexpected ways.

Notes

1 See J. Simpson, Live and Life in Virtual Theatre: Adapting Traditional Theatre Processes to Engage Creatives in Digital Immersive Technologies, *Journal: Electronic Workshops in Computing*, 2021.
2 G. Genette, *The Aesthetic Relation*, Cornell University Press, Ithaca, NY, 2010.
3 E. Dissanayake, *Homo Aestheticus: Where Art Comes From and Why*, University of Washington Press, Seattle, 1992.
4 See A. Berleant, *Aesthetics beyond the Arts: New and Recent Essays*, Ashgate, Farnham, 2013.
5 See H. Hepp, *Deep Mediatization*, Routledge, New York, 2020.
6 See V. Turner, *The Ritual Process: Structure and Anti-Structure*, Aldine, Chicago, IL, 1969.
7 The myth of the lamella is a concept developed by the French psychoanalyst Jacques Lacan, building on the psychoanalytic theories of Sigmund Freud. The lamella represents a kind of imaginary organ that embodies the vital and desiring energy persisting beyond physical death. In detail, the lamella is described as an immortal entity, a formless energy that survives independently from the physical body. It serves as a powerful psychoanalytic symbol exploring the untamable and inexhaustible nature of human desire, highlighting how it is a perpetual and immutable force transcending physical life. See J. Lacan, *La posizione dell'inconscio*, in *Scritti*, Einaudi, Torino, 1966.
8 M. McLuhan, *Understanding Media: The Extensions of Man*, McGraw Hill, New York, 1964.
9 G. Deleuze, *Essays Critical and Clinical*, University of Minnesota Press, Minneapolis, 1997.
10 L. Manovich, *The Language of New Media*, MIT Press, Cambridge, MA, 2001.
11 Cie Gilles Jobin brings together groups of digital and movement artists with the aim of participating in the creation of an international-level local creative digital ecosystem. Their goal is to expand knowledge and increase digital experimentation and real-time remote collaboration on digital and hybrid projects from the perspective of live performance. See https://www.gillesjobin.com/.
12 Direction and Choreography: Gilles Jobin. Performers: Susana Panadés Diaz (Lead Dancer), Rudi van der Merwe, Jozsef Trefeli, 3D Artist: Tristan Siodlak, see https://www.gillesjobin.com/en/creation/cosmogonie-video-installation-using-motion-capture/.
13 Virtual Crossings is an ongoing international digital project aiming at connecting simultaneously clusters of artists to collaborate remotely, on digital and

movement related art projects and research. We did three editions Geneva-BuenosAires (2020+2021); Geneva-Melbourne-Auckland (2021) Geneva at LeP-laza cinema (2022). See https://www.gillesjobin.com/en/creation/virtual-crossings/.

14 https://www.gillesjobin.com/en/studios44-mocaplab/, [accessed 26 September 2023].

15 *Artificial Intelligence in the Metaverse: Bridging the Virtual and Real*, 9th December 2021, https://www.xrtoday.com/virtual-reality/artificial-intelligence-in-the-metaverse-bridging-the-virtual-and-real/, [accessed 26 September 2023].

16 E. Mircea, *Myth and Reality*, Harper & Row, New York, 1963.

17 A. Lepecki, *Exhausting Dance: Performance and the Politics of Movement*, Routledge, New York, 2006.

18 R.S. Sugirtharajah, *Postcolonial Reconfigurations: An Alternative Way of Reading the Bible and Doing Theology*, Chalice Press, St. Louis, 2003, pp. 23–24.

19 E. Castronova, *Synthetic Worlds: The Business and Culture of Online Games*, University of Chicago Press, 2005.

20 https://xrmust.com/xrmagazine/gilles-jobin-virtual-crossings/, [accessed 26 September 2023].

21 J. Rancière, The Emancipated Spectator, Verso Books, New York, 2011.

22 A. Pietrobon, *Interview to Gilles Jobin, "Working in real time means having a piece that is always evolving"*, https://www.gillesjobin.com/wp-content/uploads/2022/02/cosmogony_16_02_22_xrmust.pdf, 16th February 2022, [accessed 26 September 2023].

23 A.D. Aczel, *Entanglement: The Greatest Mistery of Physics*, Four Walls Eight Windows, New York, 2001.

24 Publio Ovidio Nasone, Metamorfosi, in A.D. Aczel, *Entanglement: The Greatest Mistery of Physics*, Four Walls Eight Windows, New York, 2001, *Opere*, Biblioteca della Pléiade, Einaudi, Torino, 2000.

25 https://xrmust.com/xrmagazine/gilles-jobin-virtual-crossings/.

26 See M.B.N. Hansen, *Body in code: Interfaces with New Media*, Routledge, New York and Oxford, 2006.

27 See A. Pietrobon, *Interview to Gilles Jobin, "Working in Real Time Means Having a Piece that Is Always Evolving" - Gilles Jobin (COSMOGONY)*, https://www.gillesjobin.com/wp-content/uploads/2022/02/cosmogony_16_02_22_xrmust.pdf, [accessed 16 February 2022].

28 S. Hagermann, I. Pluta, Quels rôles pour le spectateur à l'ère numérique? , Épistémé, 2023, pp. 75–76.

29 S. Hagermann, I. Pluta, Quels rôles pour le spectateur à l'ère numérique ? , Épistémé, 2023 p. 76. See also. E. Pastor, *L'interprète au sein du dispositif technologique. Une humanité augmentée*, in Izabella Pluta (dir.), *Scènes numériques/Digital Stages*, Presses universitaires de Rennes, Rennes, 2022, pp. 275–290.

30 See N. Barile, *Communication in The Hybrid Onthologies*, Bocconi University Press, Milano, 2022, pp. 109–117.

31 See E. Quinz, *Il cerchio invisibile*, Mimesis, Milano, 2014.

32 R. Diodato, *The Sensible Invisible. Itineraries in Aesthetic Ontology*, Mimesis International, Milano, 2015, p. 33.

33 C. Geertz, *The Interpretation of Cultures*, Basic Books, New York, 1973, p. 88.

34 See D. Ihde, *Experimental Phenomenology. An Introduction, Albany*, University of New York Press, New York, 2012.

35 See M. Mauss, *Sociologie et anthropologie*, PUF, Parigi, 1950.

36 M. Canevari, *Lo specchio infedele. Prospettive per il paradigma teatrale in antropologia*, Mimesis, Milano, 2015, p. 45.

37 Cosmogony. Explore New Territories, Techvangart, 27th January 2022, https://www.gillesjobin.com/wp-content/uploads/2022/02/cosmogony_27_01_22_techvangart.pdf, [accessed 20 March 2024].
38 L. Jarvis, *Immersive Embodiment: Theatres of Mislocalized Sensation*, Palgrave Macmillan, London, 2019, pp. 73–95.
39 https://xrmust.com/xrmagazine/gilles-jobin-virtual-crossings/, [accessed 20 March 2024].
40 P. Auslander, Digital Liveness: A Historico-Philosophical Perspective, in *PAJ: A Journal of Performance and Art*, 34(3), pp. 3–11, 2012.
41 See G. Deleuze, *L'actuel et le virtuel*, Flammarion, 1995, p. 179 ss.
42 See M.B.N. Hansen, *Body in Code: Interfaces with New Media*, Routledge, New York and Oxford, 2006.
43 http://www.simonsenn.com/be-arielle-f/, [accessed 24 September 2023].
44 Text extracted from the video of the performance kindly provided by the artist, min.1–3.45.
45 Text extracted from the video of the performance kindly provided by the artist, min.31.00.
46 The study of psychosocial phenomena such as the Proteus effect is part of a broader field of research that examines the behavior of individuals engaged in Computer-Mediated Communication (CMC). See B. Sherrick, J. Howe, T. Waddell, The Role of Stereotypical Beliefs in Gender-Based Activation of the Proteus Effect, in *Computers in Human Behavior*, 38, 2014.
47 B. Sherrick, J. Howe, T.Waddell,*The Role of Stereotypical Beliefs in Gender-Based Activation of the Proteus Effect*, in *Computers in Human Behavior*, 38, 2014.
48 See D. Valentine, *Imagining Transgender: An Ethnography of a Category*, Duke University Press, Durham, NC, 2007.
49 L. Jarvis, *Immersive Embodiment: Theatres of Mislocalized Sensation*, Palgrave Macmillan, London, 2019, pp. 99–144.
50 E. Fuoco, *Né qui, né ora: peripezie mediali della performance contemporanea*, Ledizioni, Milano, 2022, pp. 56–65.
51 R. Diodato, *The sensible invisible. Itineraries in Aesthetic Ontology*, Mimesis International, Milano, 2015.
52 S. Turkle, *Life on the Screen: Identity in the Age of the Internet*, Simon & Schuster, New York, 1995.
53 When virtual experience was not yet sophisticatedly 'embodied'.
54 J. Baudrillard, *Simulacra and Simulation*, University of Michigan Press, Ann Arbor, 1981.
55 W.W. Ryszard, Hyperreality and Simulacrum: Jean Baudrillar d and European Postmodernism, in *European Journal of Interdisciplinary Studies*, May–August 2017, 76–80, doi: https://doi.org/10.26417/ejis.v8i1.
56 M. Crossley, *Intermedial Theatre. Principle and Practice*, Red Globe Press, London, 2019.
57 U. Eco, *Faith in Fakes – Travels in Hyperreality,* Vintage, London, 1998.
58 N. Goodman, *Ways of Worldmaking*, Hackett Publishing Company, Indianapolis, 1976, p. 126.
59 N. Goodman, *Ways of Worldmaking*, Hackett Publishing Company, Indianapolis, 1976, p. 22.
60 N. Goodman, *Of Mind and Other Matters*, Harvard University Press, Cambridge, MA, 1984, p. 45.
61 See F.A. Kittler, *Gramophone, Film, Typewriter*, Stanford University Press, Stanford, CA, 1999.

62 Aristotle, IV sec.a.C., I, 4, 1253 b, 30–35.
63 Aristotle, IV sec.a.C., I, 4, 1253 b, 30–35.
64 P. Virilio, *The Aesthetics of Disappearance*, MIT Press, Cambridge, MA, 1991.
65 See S. Dixon, *Cybernetic-Existentialism Freedom, Systems, and Being-for-Others in Contemporary Arts and Performance*, Routledge, 2020.
66 H.-T. Lehmann, *Postdramatisches Theater*, Frankfurt a. M., Verlag der Autoren, 1999, pp. 254–260.
67 J.-L. Nancy, *Corpus*, Editions Métailié, Paris, 1992.
68 The idea has been formulated in a variety of different terms, such as the "society of the spectacle" (G. Debord [1967], *Society of the Spectacle*, Black&Red, 2002) and incorporated almost everywhere into "performative society" (B. Kershaw, Dramas of the Performative Society: Theatre at the End of its Tether, in *New Theatre Quarterly*,17(3), 203–211, 2001, doi:10.1017/S0266464X0001472X), the "spectacularisation of culture" (U. Eco [1985], Sugli specchi e altri saggi. Il segno, la rappresentazione, l'illusione, l'immagine, Bompiani, Milano, 2001). See also Edited by S. Bay-Cheng, C. Kattenbelt et al., *Mapping Intermediality in Performance*, Amsterdam University Press, 2020.
69 S. Bay-Cheng, C. Kattenbelt et al., *Mapping Intermediality in Performance*, Amsterdam University Press, 2020, p. 34.
70 See L. Groot Nibbelink, S. Merx, Presence and Perception: Analysing Intermediality in Performance, in S. Bay-Cheng, C. Kattenbelt, Chiel, Lavender et al. (eds), *Mapping Intermediality in Performance*, Amsterdam University Press, 2020, pp. 218–229.
71 See G. Giannachi, *Virtual Theatre: An Introduction*, Routledge, London, 2004.
72 See L. Groot Nibbelink, S. Merx, *Presence and Perception: Analysing Intermediality* in Performance, in S. Bay-Cheng, C. Kattenbelt, Chiel, Lavender et al. (eds), *Mapping Intermediality in Performance*, Amsterdam University Press, 2020; M. Causey, *Theatre and Performance in Digital Culture: From Simulation to Embeddedness*, Routledge, New York, 2006.
73 N. Borriaud, L'esthétique relayionnelle, Les presses du reel, Dijon, 1998, p. 15.
74 N. Borriaud, L'esthétique relayionnelle, Les presses du reel, Dijon, 1998, p. 141.
75 The traditional distinction between onsite and online [Turkle 1995] no longer seems adequate to describe the complexity of ongoing transformations. As suggested by Turkle in the 1990s and further developed by Floridi in 2015 with the concept of onlife, the continuum between online/on-life/off-life. [Floridi 2015]. S. Turkle, *Life on the Screen: Identity in the Age of the Internet*, Simon & Schuster, New York, 1995, see also L. Bollini, Space as a Narrative Interface. Phygital Interactive Storytelling in the Field of Cultural Heritage, in *Multidisciplinary Aspects of Design*, Springer, Cham, 2024, pp. 613–622, issn978-3-031–49810-7, doi 10.1007/978-3-031–49811-4_58. And L. Floridi, *The Onlife Manifesto: Being Human in a Hyperconnected Era*, Springer, Cham, 2015.
76 L. Bollini, Space as a Narrative Interface. Phygital Interactive Storytelling in the Field of Cultural Heritage, in *Multidisciplinary Aspects of Design*, Springer, Cham, 2024.
77 M. Slater, V. Linakis, M. Usoh, R. Kooper, Immersion, Presence, and Performance in Virtual Environments: An Experiment with Tri-dimensional Chess, in *ACM Symposium on Virtual Reality Software and Technology (VRST)*, 1999, pp. 163–172.
78 K. Kwastek, *Aesthetics of Interaction in Digital Art*, MIT Press, Cambridge, MA, 2013.
79 The Oxford English Dictionary defines the metaverse as "a virtual-reality space in which users can interact with a computer-generated environment and other users", https://www.oed.com/dictionary/metaverse_n?tl=true.
80 K. Hyeonyeong, L. Hwansoo, Performing Arts Metaverse: The Effect of Perceived Distance and Subjective Experience, in *Computers in Human Behavior*, 146, September 2023, https://doi.org/10.1016/j.chb.2023.107827.

81 See M. Ball, *The Metaverse: And How It Will Revolutionize Everything*, W. W. Norton & Company, 2022.
82 See S. Dixon, *Digital Performance: A History of New Media in Theater, Dance, Performance Art, and Installation*, The MIT Press, Cambridge, MA, 2007.
83 E. Fischer-Lichte, *The Transformative Power of Performance: A New Aesthetics*, Routledge, London and New York, 2008, p. 177.
84 E. Quinz, Il cerchio invisibile, Mimesis, Milano, 2014, p. 11 ss.
85 C. Biasucci, M.E. Drumwright, *Behavioral Ethics in Practice: Why We Sometimes Make the Wrong Decisions*, Routledge, New York, 2019.
86 See E. Goffman, *Frame Analysis: An Essay on the Organization of Experience*, Harvard University Press, Cambridge, MA, 1974, p. 32.
87 See E. Goffman, *Frame Analysis: An Essay on the Organization of Experience*, Harvard University Press, Cambridge, MA, 1974, p. 32.
88 *Le Bal de Paris* de Blanca Li was awarded the Lion for Best VR Experience at the 2021 Venice Film Festival, 78ª Mostra Internazionale d'Arte Cinematografica di Venezia, https://www.blancali.com/en/spectacle/le-bal-de-paris-de-blanca-li-en/.
89 See L. Meloni, D. Boriani, *Arte cinetica, programmata, interattiva*, Manfredi Edizioni, Imola, 2018.
90 Blanca Li's choreography is a fundamental element of the show, combining elegant and dynamic movements that reflect the energy and spirit of the Parisian dance. The dancers, including Susana Panadés Diaz, Rudi van der Merwe, and Jozsef Trefeli, perform complex and engaging numbers that enrich the narrative and the immersive experience.
91 Blanca Li's "Le Bal de Paris" utilizes VR headsets and technology developed by Black Light Studio. The performance incorporates a full VR suite with headsets and tracking devices for hands and legs, allowing participants to fully immerse themselves in a virtual ballroom experience. The VR experience is supported by HTC Vive (official partner for virtual reality) and involves high levels of interaction and visual fidelity, including custom avatars and costumes designed in collaboration with Chanel (Belgrade Dance Festival 2023).
92 https://belgradedancefestival.com/en/performance/le-bal-de-paris.
93 https://www.festivaldispoleto.com/eventi/le-bal-de-paris.
94 F. Nietzsche (1886), *Beyond Good and Evil. Prelude to a Philosophy of the Future*, Cambridge University Press, 2001.
95 N. Yee, J.N. Bailenson, The Proteus Effect: The Effect of Transformed Self-Representation on Behavior, in *Human Communication Research*, 33(3), 271–290, 2007.
96 L. Charmaine, *Fun starting point for virtual-reality newbies in Le Bal De Paris De Blanca Li*, https://www.straitstimes.com/life/arts/dance-review-fun-starting-point-for-virtual-reality-newbies-in-le-bal-de-paris-de-blanca-li, 8th March 2024. See also A. Rostami, D. McMillan, The Normal Natural Troubles of Virtual Reality in Mixed-Reality Performances, in *CHI Conference on Human Factors in Computing Systems*, New Orleans, LA, April 2022, pp. 1–22, doi:10.1145/3491102.3502139.
97 See E. Goffman, *The Presentation of Self in Everyday Life*, Anchor Books, London, 1959, and V. Turner, *From Ritual to Theatre: The Human Seriousness of Play*, Performing Arts Journal Publications, 1982.
98 S. Bay-Cheng, C. Kattenbelt et al., *Mapping Intermediality in Performance*, Amsterdam University Press, 2010, p. 32.
99 See M. Seel, *The Art of Separation: On the Concept of Aesthetic Rationality*, Suhrkamp Verlag, Frankfurt am Main, 1985, p. 127 and 247.
100 E. Goffman, *Frame Analysis: An Essay on the Organization of Experience*, Harvard University Press, Cambridge, MA, 1974.

101 E. Goffman, *Frame Analysis: An Essay on the Organization of Experience*, Harvard University Press, Cambridge, MA, 1974.
102 E. Goffman, *Frame Analysis: An Essay on the Organization of Experience*, Harvard University Press, Cambridge, MA, 1974.
103 C. Kihm, *Le spectateur expérimenté*, in E. During, L. Jeanpierre, C. Kihm, D. Zabunyan (eds), in *Actu, De l'expérimental dans l'art*, Les presses du reel, Dijon, 2008, p. 345, see also E. Quinz, Il cerchio invisibile, Mimesis, Milano, 2014, pp. 123–137.
104 See E. Hornung, *Conceptions of God in Ancient Egypt: The One and the Many*, Cornell University Press, New York, 1996.
105 See P. Descola, *Beyond Nature and Culture*, University of Chicago Press, Chicago, IL, 2013.

3 Artistic Experimentation and the Role of Performing Robots

This chapter examines the integration of advanced technologies in live theatrical performances, including pre-recorded visual and audio elements, interactive scenography, and motion capture technology. It questions the nature of live theater in the context of digital experiences, emphasizing the importance of "live" modifications by the design team. By adopting an aesthetic and anthropological perspective, the analysis explores the interplay between art and technology, highlighting their mutual influence on human cognition and culture. It addresses the transformation of theatrical experiences through virtual reality and other immersive technologies, redefining audience engagement and the performer's role in a mediated cultural landscape. The chapter underscores the evolving relationship between real and virtual environments, fostering new forms of interaction and artistic expression.

3.1 Machines "Like" Me: Rimini Protokoll's Uncanny Valley

Robots were introduced as characters in dramas in the 20th century through Karel Čapek's work, *R.U.R.* (*Rossum's Universal Robots*), first published in 1920 and staged in 1921. This play is noteworthy not only for its innovative theme but also because it introduced the term "robot", derived from the Czech word *robota* meaning forced labor or servitude. The main characters are artificial beings constructed from synthetic biological material, indistinguishable from humans in appearance and capable of performing all human actions. If one were to trace a genealogical excursus of the notion of robots, it would become evident that from the outset, robotic agents were conceived as artifacts whose primary function is to work for us.[1] In the transition from the science fiction scenario depicted in Čapek's play to the current technological world, robots have remained artificial agents that autonomously perform certain tasks on our behalf.

Despite the initial impact of *R.U.R.* it took decades before robotic characters became a regular presence in Western theaters.[2] It was only in the latter half of the century that works such as Alan Ayckbourn's *Henceforward*

DOI: 10.4324/9781032676937-4

(1987), *Comic Potential* (1999), and *Surprises* (2012), along with lighter and humorous theatrical and musical sci-fi works like *Return to the Forbidden Planet*, began to explore robot themes more frequently.[3]

In this manner, it will undoubtedly be possible to emancipate robotic agents from the singular alternative that has rendered them either perfectly faithful servants or slaves ready to rebel and thereby combat humans. These experimental artistic performances[4] indeed approximate the concept of robots as companions in learning and as support in the processes of self-awareness and human development, largely conceived by Eastern cultures, particularly Japanese culture, where autonomous robots have long existed. This scenario, associated with manga, animated cartoons, and fantasy literature, is viewed in a positive and creative light.

The practice of using real robots to represent themselves on stage is a much more recent phenomenon, emerging as a trend only with the advancement of robotic technologies in the 21st century. Indeed, the new millennium has witnessed a growing interest among playwrights and theatrical creators in the socio-historical and cultural implications of robots. These artists have begun to experiment with the use of real robots on stage, exploring interactions between humans and machines, and reflecting on the ethical and philosophical challenges posed by coexistence with advanced technologies. This shift reflects a broader transformation in the perception of robots, moving from mere figures of narrative to entities with a tangible and significant impact on human society and culture.[5]

Robot mythology is an immersive, co-evolutionary narrative in which humans are included, without defined boundaries. It is something already strictly related to our culture and historical time: "Artificial Intelligence and robots could be seen as an emergent species".[6] The "sense of other", in artistic and more in anthropological terms, is strongly reformulated in both of its meanings. With regard to having a sense of family, direction, or rhythm, it can be understood as an innate or acquired gift of this kind. In its second meaning, it refers to a "sense of others" or an understanding of what has meaning for others.[7]

Adrienne Mayor, in her book *Gods and Robots: Myths, Machines, and Ancient Dreams of Technology*,[8] draws a fascinating and insightful connection between the myth of Pygmalion and artificial intelligence (AI). In the myth of Pygmalion, the sculptor falls in love with a statue he has carved, and the goddess Aphrodite grants his wish by transforming the statue into a living woman. Mayor sees in this story a clear analogy to modern developments in AI and robotics. According to Mayor, the myth of Pygmalion reflects the enduring human desire to create artificial life and the complex emotions associated with this process. "The dream of animating the inanimate has been part of human consciousness for millennia",[9] and the aspiration to give life to inanimate objects is not just a product of the modern technological era but has deep roots in our collective imagination. And, we could consider this encounter, between non-human actor and human spectator, a formative and theatrical event and study it as such, with a specific focus on process and processual qualities.

The theater historian Jean Alter, in his text *Sociosemiotic Theory of Theater*[10] identifies a typical function of theatrical action that he calls *performant*. A *performant* act consists of showing one's exceptional skills and virtuosity regardless of the referential intent of telling a story. In such terms, a performance is to be understood as an act of "extraordinary" expressiveness that creates interest and gathers an audience around it precisely because it is not comparable to any common acts. This is the most suitable context for describing the creation of the Rimini Protokoll[11] theater collective, which will be discussed later, an example whose anti-structural performative character is evident, centered on explicating and staging the latent cultural codes of the system and thus inducing self-reflection.

On the contemporary stage, we participate in transmedia narratives[12] that, by moving between different types of media, help to implement or integrate the spectator's experience with distinct and unprecedented information. The body, understood as flesh-and-blood human, undergoes a process of remediation[13] and democratization from its biological and material being and, paradoxically, is summoned into its dematerialization, dehumanization and, even more provocatively, absence.

> Robotic and virtual figures have become increasingly visible in live performance, functioning more as actors than simply as objects, props, or decor. Dramaturgically they link new perspectives on technology, media, politics, and ideas of subjectivity and existence within hybrid performance events that blur the traditional theatrical borders between live and mediated effects.[14]

Robotic and virtual figures have become increasingly visible in live performances, functioning more as actors rather than merely objects, accessories, or decorations.[15] Reflecting on these new hybrid artistic practices can only be interdisciplinary and develop not only in terms of aesthetics but also in philosophy and anthropology.

The relationship established since the start of the 21st century between humanity, technology and the environment determines the development of a vast phenomenology of occurrences capable of creating new development trajectories for individuals and society. Especially fundamental changes, such as those stemming from the inclusion of robots[16]—in terms of complex technical objects—not only in support of various daily activities but by now active in creative and cultural contexts, have determined a rearrangement of the paradigmatic frameworks within which the dynamics of construction of human knowledge and, consequently, the prospects for research and artistic creation, are delineated.[17] The most contemporary artists seem to manifest the desire to outstrip a "humanism narcissism", no longer taking for granted the place and definition of human; we no longer start from the human and the world, but the human as an integrated part, together with other entities in the world. Robots, machines, must be considered as "performing objects".

This 'performing object' concept is commonly used by scholars and practitioners of puppetry to designate stage props, such as puppets or masks for performance. In addition, the puppeteer and scholar John Bell includes props such as sculptures, paintings, and ritual objects in performance, and refers to the 18th- and 19th-century automatons. Bell also discusses how gadgets such as smartphones or televisions 'perform with us and for us', by transmitting 'stories and ideas'.[18] This broadened notion of performing props highlights the significance of multi-layered relationships with everyday objects, and the fact that these objects seem to have an 'ability' to engage with people. These performing and performative objects are increasingly protagonists of a post-human theater, a place of multiple interactions between the human and the machine.

The initial performance selected for this analysis is the production by the Rimini Protokoll[19] collective, *Uncanny Valley*[20] in which it is the spectator who is the real actor embodying the performative experience of the monologue recited by the robot and gives it meaning, according to his embodied experience.[21]

> Uncanny Valley begins when the audience enters the space of performance. At that moment, the audience sees a white screen on the left of a square carpet and on the right, there is an armchair on which a humanoid robot is sitting next to a laptop on a small table. When everyone is seated, the stage is bathed in light. The humanoid introduces itself: «I am Thomas Melle». He tells us that we are here for his lecture on instability and the "Uncanny Valley" phenomenon[22]. A few photos of his younger days appear on the screen, which illustrate the story of what type of child he was and his questions as to his identity during his adolescence. In 2016, he published the book Die Welt im Rücken (The World at your Back) in which he talks about his illness and gives his definition of this psychological dysfunction. [...] Melle confides in the performance while being absent on stage. He is represented by a robot that embodies his absence. Melle only appears on stage as a video image, but gives his voice to the dubbed robot, especially in the German version.[23]

Should the choice of placing an anthropomorphized robot at the center of a theatrical stage annoy or amaze? The answer to date is certainly no, we know from well-established studies that humans tend to "humanize" any concept or entity by a natural cognitive process of acquisition and understanding of things and that the best interface of a human being is something that looks human. What can create at least a sense of estrangement is the choice of building a robotic twin—a Geminoid[24]—of an existing and known human being. The human appearance of this robot actor is as astonishing as is frustrating its inability to interact outside the given programming or budge from the chair where it sits throughout the hour-long play. It is like a robotic puppet, unconnected to any AI system, devoid of autonomy or creative ability (Figure 3.1).

Isn't it perhaps disruptive to see how this cultural trend, this social posture that invades even an art field such as the theater, a spectacle that is by its very nature linked to, produced and staged by human beings? So, it comes naturally to ask an almost Kafkaesque question: "Whom does theater speak to? Whom do the characters speak to?" A human character, a well-known German writer, is presented on stage through a video and replicated by an android, a humanoid robot whose exposed circuitry on the back of its head shows the audience that it is actually a fake and not a 'real' actor. Talking about his human life, he confesses how he has sometimes felt outside his body, like an external observer of himself: "How authentic are your memories? To overcome my own panic, I act like a machine". (he/it said). He reflects with the public on what mass-media phenomena can cause: an individual discomfort in relating to the public, never in the first person, but always "mediated", to the point of believing that the authenticity of one's own person is in that image of oneself which is transmitted. And it continues:

> How are you? Right here, right now…? And how do you feel having sat here and listened to me? [If you've come to see an actor, you're in the wrong place. But if you've come to see something authentic, you're also in the wrong place. […] Because I don't want to expose myself, not as the "real thing", the "real person", because… look: the stage enlarges everything to an unbearable degree?.

Figure 3.1 Uncanny Valley - directed by Stefan Kaegi. Photo @ Gabriela Neeb

The android recounts with ironic gibes the life of the scientist Alan Turing, known as one of the pioneers of the study of computer logic and as one of the first scientists to take an interest in the subject of AI. The robotic actor continues:

> Are humans not numbers of algorithms that must necessarily position themselves, to relate to others? Aren't machines capable of being as perfect as humans, or is it simply that there is no concept of "error" because everything is linked to cause and effect?

Not all artistic practices that deal with technology or that use living beings must be considered "post-human material". In a certain sense, making post-human art also implies being (a cluster of) post-human subjectivities, understanding and consciously exercising certain ethics that include not only the supplanting of anthropocentric conceptions in relations with the living, the semi-living and the inorganic, but also the ability to conceive new methodologies of thought and practice[25]. In this sense, *Uncanny Valley*,[26] whose only protagonist is a humanoid robot, provides the insight for arguing how contemporary theatre, in its exasperated experimentation, has attempted to express the idea of humanness in terms of processualism, as a "becoming", a changing entity that can only be provisionally defined on the basis of its relations with the other,[27] so as to make the post-human body, whether exhibited or absent, appear as a new stage in the ongoing process of cultural hybridization, as a new frontier of our complex relationship with the world.[28]

As a Frankenstein, toward the middle of the performance, the robot begins to show films and photos of its own creation, as it assembles its being from the silicone cast of Melle's face or the 3D reconstruction of his limbs; that "thing" which, looking at the audience, begins to ask: "You find me uncanny? [...] I have no idea who you are. But do you know? [...] Maybe something about me may offend you". But robots, which are increasingly anthropomorphic, arouse unease and disorientation, though are they not perhaps born from our desire for immortality, assisted longevity or even divine omnipotence, capable of creation? Indeed, the vision of the android, which is explicitly expressed in the final part of the performance, is that of a fallible machine, whose control is still, despite its sophisticated human appearance, in human hands. A machine that has its weaknesses, bugs and a sort of technological singularity. It is no mirror symmetry or claim to autonomy that is brought on stage but a reflection on difference, otherness, what distinguishes us as human beings from "machines" and it is done in an exemplary place of the multiple facets of human complexity: a theater stage.

What happens to the concept of being-for-others, closely related to the theatrical world, if the only on-stage character is an android? Apropos of that inter-corporal relationship that anthropologist Victor Turner spoke of configured as the coexistence of performer and bystander[29] thus making it possible to recreate those "original" relational conditions in the mind of the second, to produce an interaction based on recognition and 'immediacy? To attempt an answer, we must start with new expressions, revolutionary connections. In fact, to understand the process that this type of performance triggers in the spectator, in our opinion it is more apt to speak of intra-action.[30]

Building upon a concept of technology based on New Materialisms, which requires

> "an understanding of the nature of the relationship between discursive practices and material phenomena, an accounting of "non-human" as well as "human" forms of agency, and an understanding of the precise causal nature of productive practices that takes account of the fullness of matter's implication in its ongoing historicity.[31]

Karen Barad speaks of post-humanism using some assumptions of quantum physics, and we refer to her theory of agential realism, by which the universe comprises phenomena that are "the ontological inseparability of intra-agent" and non-interacting agencies. Intra-action is a neologism that Barad introduces. For her, "phenomena or objects do not precede their interaction; rather, 'objects' emerge through particular intra-actions".[32] Thus, the devices that produce phenomena are not assemblages of humans and non-humans—as in actor-network theory). Rather, they are the condition of possibility of 'humans' and 'non-humans', not simply as ideational concepts, but in their materiality. Hence, in the case of a performance that integrates the non-human, a robot, a machine into its process, we could state that an intra-action phenomenon and an entanglement phenomenon are created.

Barad's theory seems to be a fruitful trajectory for analyzing a post-human performance as a choice in this analysis whose dichotomous-hierarchical polarizations between outer/inner and human/non-human disappear.[33] If we adopt the perspective of quantum physics, at whose base we find an intrinsic ontological indeterminacy, a constitutive ontological differentiation of all entities of and in the world, not based on anthropocentrism, we can understand what kind of experience can exist between a robotic actor and a human spectator, by probing the material-discursive practices of construction and reification of borders with the "non-human" in an artistic field centered—from its origins—on ritual and relationship.

Theater means dealing with an era of forms, overturning traditional axioms, and subverting historical analytic paradigms. The body, at the center of the live performing arts, becomes itself a form and a trans-conductor, and to do so, it betrays its own human substance so that it can acquire complex and unprecedented levels of expressive technique. The total assimilation of new technologies, from AI to robotics, certainly has more to do with what we define as *performance*, in the broad sense that Schechner gives to the term,[34] than with theater as such, the theatron,[35] the place for seeing, co-presence and relational reciprocity.

A classical, frontal drama structure with passive spectators is an apparent anomaly in the Rimini Protokoll's interactive concept. But using the format of a traditional theatre to subvert it from the inside with a robotic actor is perfectly in keeping with the German company's provocative, desecrating spirit. Placing this mechanism before an apparently passive audience means putting their backs to the wall, cracking and questioning the established functions of theatre, creating bewilderment and embarrassment when they must consider whether or not it is appropriate to applaud at the end of the performance. An expedient for trying to stimulate reflections on how weak and at the same time powerful the confrontation can appear, on a stage, between man and robot. A provocation, it would seem, to direct, to rethink identity itself and the performative ways of constructing the human. We are not confronted with an object, an entity/subject that pre-exists, an element external to a world in which it subsequently enters and interacts, but with an iterated becoming, a robotic "entity" that occurs in its own intra-acting, connected with the audience that watches it.

To develop this line of reasoning, we realize that, more than a problem of understanding a phenomenon, it is a problem of language: there are no new words and new categories to describe the new performative entities created by this fourth industrial revolution.[36] What is left of theatre in a performance acted by a single android? Maybe the collective experience of sharing the enjoyment of the play, the aesthetic, synesthetic, empathic dimension of what we feel in relation to what the human person sitting next to us or in front of us or behind us feels or doesn't feel. But we can think that the term "theatrical communication" is supplanted by that of "connection", or does the mere hypothesis of artificial life, and of the robot as its symbol invading the theatrical stage, lay the foundations for an opposite ontological condition, based on separation, estrangement, or incommunicability? These are reflections arising from the boundary that the non-human, on stage as the sole protagonist, can trigger, specifically as a performance which shows it as unstable by virtue of the absence of the main "concept of being-for others", closely related to the theatrical world, due to the total absence of human on stage?

In the last scene of the performance, we see the image of the "real" Melle appear on the blackboard, who from the video ponders his virtual presence: "I'm here and not here"; he then directs motion commands to what he defines as his puppet and continues: "My robot is my Dorian Gray and I'm the decaying image". He commands that the plug be pulled to show his power over the true protagonist of the entire play, then turns it back on, until he reaches his last "stop" leading to the darkness at the end of the play. In artistic terms, it is certainly interesting to see how the disturbing effect can become a conscious and not necessarily unpleasant tool.

> It is important to note how pure art avoids to a fair extent the absolute and complete imitation of nature and of the living being, knowing full well that some people are prone to get uneasy, generating the uncanny can be attempted in true art but solely and exclusively for artistic intentions and meanings.[37]

To which "myth" does this contemporary performative experimentation lead us? Exploring its meaning, detaching itself from written dramaturgy, we certainly find in some passages a theatrical actualization of conflict, linked to Heidegger's philosophy. In *The Question Concerning Technology*,[38] Martin Heidegger explores modern technology not just as a tool but as a way of revealing the world that conditions our understanding of reality.[39] He introduces the concept of *Ge-stell* (enframing), describing how everything is set in position and made available for use.[40] Heidegger warns that this mode of revealing can obscure other more authentic forms of relationship with being, pushing toward a utilitarian and reductive worldview.[41] Technology, therefore, is not merely a means but a mode of existence that profoundly transforms the relationship between humans, the world, and being.[42] Heidegger's idea of the origin of conflict can be connected and integrated into this context, as he sees conflict as an intrinsic element of the human condition and the way humans relate to being and truth.[43] Conflict arises from the misalignment between the mode of revealing imposed by technology and more genuine forms of existence. This misalignment creates tension between the technical-instrumental domain and the need for a more authentic and less alienated relationship with being.[44] In this sense, technology, in its essence, not only changes the way we see and interact with the world but also introduces an ontological conflict, as the dominance of *Ge-stell* threatens to reduce humans to mere functionaries of technology, distancing them from their true essence and potential for authentic existence.[45] This conflict is thus fundamental for understanding how humans can rediscover and reaffirm their essential relationship with being, beyond the limitations imposed by the technical view of the world. Conflict, if approached with a problem-solving orientation, has an evolutionary role, fostering growth and transformation of relationships in a positive sense.

In this case, it can: prevent relationship stagnation; stimulate interest and curiosity; address problems that might otherwise remain unresolved and find new solutions to old problems. We can see how fears associated with dystopian and apocalyptic narratives inherited from the Western collective imagination related to science fiction and the cinematic world find an apparent resolution in the 'theatrical' territory. The theater, always a place of catharsis[46] and reflection,[47] becomes the place where humans materially experience guidance from a mechanical and synthetic agent and measure how and to what extent they can train it in turn. The degree of autonomy and creativity is followed, also proceeding through clashes that generate glitches, like the "fictitious" one shown on stage by the robot pretending to malfunction, losing control of itself during a quarrel with the "real" Melle who appears to us on video.

The glitch can serve as a potent creative resource in technological performance by leveraging the aesthetics of error and disruption to challenge and expand traditional narratives and visual expectations.[48] Moreover, the incorporation of glitches can serve as a form of critique and reflexivity within the performance, highlighting the fragility and fallibility of technological systems. This can prompt audiences to question the reliability and control of digital environments, fostering a deeper understanding of the interaction between humans and technology.[49] By embracing the glitch, performers can subvert the polished veneer of digital perfection, instead celebrating the raw and imperfect nature of technological interactions. This not only democratizes the creative process, making it more accessible and relatable, but also aligns with broader postmodern and post-digital discourses that value fragmentation, hybridity, and the disruption of linear narratives.[50] Thus, the glitch becomes a versatile tool that enriches the artistic vocabulary, allowing for innovative explorations of form, content, and meaning in the context of technological performance.

A contemporary example of this theory can be observed in object theater that uses robots as advanced puppets. In these performances, robots are not just physical objects, but they act as intentional signs that transcend their technological nature to represent an entire class of performative entities. This type of theater perfectly illustrates Eco's concept of "ostension" as robots, through their actions and interactions, cease to be mere machines and become symbols of human behaviors and cultural identities.[51] Their presence on stage becomes a performative act that explores the complexity of human-machine relationships, creating a possible world that investigates the posthuman condition and the nature of agency.

3.2 Exploring Human-Robot Interactions through Dance

In theater, the character of the robot enjoys a long history, connected to the primordial exploration of alterity and the relationship between the animate and the inanimate, the organic and the inorganic, and especially to the quintessential theatrical objects: the mask and the puppet.[52] The metaphor of the automaton or the marionette,[53] of the machine as a system devoid of the mutable intentionality of humans, is closely linked to humanity's attempt to reach the absolute through art,[54] to the desire to achieve something that transcends the contingent, the mutable, to attain that abstract and "mechanical" sublime, which harbors something divine or metaphysical.[55]

The robotics that enter the realm of art must be studied according to the undeniable influence of the context in which they interact and develop, and through an analysis that is not only quantitative but also qualitative and humanistic (aesthetic, anthropological, philosophical, and cognitive). Part of the fascination produced by robotics in the performative field is undoubtedly linked to its biomimicry and the interpretation of bio-inspired neuroscientific models that are triggered by the expression of a gesture or a facial expression that replicates human ones.[56] Contemporary theatre seems to be at the forefront of current developments in emerging new technologies; however, the literature tends to focus more on the individual impact resulting from direct interaction with robots or on technological elements[57] rather than adopting an adequate interdisciplinary approach.

As indicated by the increasing number of productions in recent years, 21st-century theater appears to be at the forefront as an experimental field for testing and implementing continuous inventions related to robotics and AI.[58] The theatrical art possesses the innate capacity to transform the inanimate into the animate, the idea into a project, and utopia into reality. Dancing means acting with one's body, being moved, creating and controlling movement to the point of forgetting one's technique, overcoming the limits of one's body, and being moved, like a puppet, by the movement itself.

Building on this premise, there is a growing pursuit of dialogue, relational interaction, and attentive engagement between human and non-human entities. Robots and dance represent an intriguing intersection between technology and artistic expression. Over the years, robots have been progressively integrated into the art world, challenging traditional notions of performance and pushing the boundaries of creativity. Robots can act as agents that amplify the effects of presence, communication, and work, in short, as collaborators. However, it is essential to ask whether they can also trigger empathetic engagement in humans.

"Empathy" whose etymology refers to what is felt inside, is a complex concept that involves mechanisms and dispositions through which an individual or an animal can "understand" the feelings of others. In the context of interpersonal relationships, empathy is distinguished from sympathy, compassion, and emotional contagion. It depends on affection, whether vague or qualified, arising from or resulting in a psychological state, and most of our emotional reactions are automatically activated in response to the bodily expression of others.

Empathy necessarily develops for a small-sized robot that resembles a child, an animal, or a woman, as it is known that culturally, people are predisposed to speak with greater care and gentleness to a woman, for example.

> Initially, it was thought that the problem could be solved through a smile or the soft and smooth texture of robots, which is why silicone was used for androids. However, this method did not prove to be the most effective. In reality, it is not possible to develop true empathy if the robot is not intelligent. Incorporating intelligence into a machine is not easy; even when connected to the Internet, a robot will still appear less intelligent than any person met on the street. How can we proceed, then? This is where the artist's role comes into play, which might seem unnecessary. To enable this capacity for empathy, it is necessary to simulate intelligence. More precisely, the form of intelligence of interest in this context is consciousness. Since we are still far from being able to simulate consciousness, the idea has been to suggest it through the staging of unconscious behaviors.[59]

Regarding the specificity of relationships with certain robots, or in general with certain animated objects, their ability to emotionally engage us is notable, despite being devoid of life. For some individuals, this can stem from the ability to identify with another's thoughts, actions, to put oneself in their place, or even to emotionally identify with another person—or object—intuitively feeling its "emotions".[60]

The issue of the ability to elicit empathy remains a fascinating and complex field of study, requiring further research and reflection. Through the selected performative examples, a key to success can be found in the relationship and dialogue between the human and non-human, enacted in a poetic and artistic dimension.

As full-fledged choreographic partners or entities, robots capture the audience's attention and pose fundamental questions in dance related to contact, listening, reactivity, and the trustful surrender of one's body. The choreographer Blanca Li, mentioned in the first chapter, sought to stage a pressing contemporary issue: the growing dependence of modern humans on machines and robots. On stage, she entrusts movement and gesture, rather than words, with her inquiries: "What are the boundaries between 'us' and 'them'?" "Can a machine replace relationships with living beings?" "Will our robotic alter egos express feelings?"

In the show *Robot*[61] (2013) by the Blanca Li Dance Company, the protagonists are small dancing objects: small-sized humanoid robots, sophisticated semi-autonomous mechanical puppets that attempt to learn and reproduce choreographic movements. This performance involves collaboration with the Japanese artistic collective Maywa Denki, known for creating automated musical instruments, and Aldebaran Robotics, creators of the small robot Nao,[62] which features prominently on stage.[63] The choice of a humanoid robot, which resembles the human body but unmistakably reveals its artificial nature, as opposed to an android robot, is certainly not coincidental.

The performance begins with a sort of "prologue" featuring a solo dancer at the center of the stage, illuminated and transformed by video projections that overlay anatomical images—from skeletal and muscular systems to the vascular system—visibly shining on his skin. As the projections intensify, the human figure begins to transform into a cyborg, evoking iconic cinematic images such as the robot from Metropolis or C-3PO from Star Wars.[64]

The choreography consists of six scenes that seem to embody Paul Valéry's thought that dance primarily represents the expression of a surplus of power. In his work *Philosophie de la danse* (1936), Valéry explains that dance emerges from the fact that humans possess more sensomotor abilities than are necessary to meet the demands of their existence. Displaying one's muscular energy and coordination is a way to channel this excess vitality and simultaneously demonstrate it to others. Valéry contrasts the vital functions of the body, linked to our survival, with those that are superfluous. Our senses, limbs, images, and signs that "command our actions and coordinate the movements of our puppet" are not only used for attacking or defending ourselves to ensure our survival. Unlike machines or animals, which perform only what is necessary and can only react to external stimuli, humans have the capacity to dance.

Valéry emphasizes that the dancer can generate, through movement, a "space-time" that fundamentally differs from that of practical life. The dancer isolates themselves in a temporal and spatial dimension of their own. The fragility exposed—by choice of the choreographer—of the robots' dance, on the other hand, lies in their dependence on the external world and the highly calibrated adjustments required. This is the difference between the training shown on stage and dance: the latter is the bodily expression of an interiority and a conduit of energy.[65]

The first scene is dedicated to the man-machine encounter, where eight contemporary dancers enter the stage, moving in synchrony with the robots. The choreography is precise and technical, highlighting the fluidity of human movements in contrast with the rigidity of the robots. This is followed by a scene depicting the "transformation of the robot" into a human being, where holographic projections and lights are used to create the illusion that the robot is becoming increasingly human, exploring the boundaries between machine and living being.

The robot becomes a simulacrum both according to the Platonic sense (from the Greek: εἴδωλον, eidolon) referring to a lower or distorted copy of reality—an imperfect and deceptive replica—and according to a more contemporary philosophy congenial to simulated realities by technological media, as proposed by Jean Baudrillard.[66] In Baudrillard's view, in postmodern society, images and signs no longer represent reality but create a new reality of simulacra that replace the real itself (Figure 3.2).

Figure 3.2 *Robot* by Blanca Lì. Photo by Ian Gavan.

In the third scene, a disembodied hand appears on stage, playing a rhythm on a drum or musical instrument. This scene plays with the idea of musical automation, demonstrating how even a separate body part can have its own life and rhythm thanks to technology. Midway through the performance, the most choreographed scene features five NAO robots in blue suits performing alongside five human dancers dressed in robot costumes reminiscent of "Lost in Space". The choreography is complex and dynamic, with dancers and robots moving together in harmony, creating a visual symphony of movement and technology. The performance *Robot* examines the relationship between robots and humans, exploring how the characteristics of each—such as a dancer's vulnerability and a robot's rigidity—can blend or merge, despite one being inanimate and the other alive.

> Designing movements for the robots, Ms. Li mentioned, is like creating an animation but with gravity; since the machines do not have the time to readjust their balance, they often fall. To address this issue, Ms. Li and her team devised a program: if a robot falls, it gets back up and catches up with the choreography.[67]

In the subsequent scene, a NAO robot dressed in a silver-blue sequined outfit and a shocking pink feather boa sings a Spanish love song in a cabaret style. This scene is playful and humorous, and I would add, it all starts with play. Play as the originating place of symbolic fiction where for the first time it establishes that 'this stands for that' and we can play the 'world of as if'.[68] Theatrical art belongs to the sociality form of play; indeed, many have argued the ludic symbolic dimension that characterizes theater, allowing the imaginary to take a concrete form of experience and body. Theatre is an imitative game of reality,[69] thus creating an unreal relationship with the imaginary, sharing human experience phenomena such as dreams or fears. Play allows us to take risks or rather to venture to the edge of the acceptable, experimenting with the unusual and the unimaginable. Situations already present in traditional theater reach new levels of intensity and distance from human action through the integration of new technologies, such as the encounter with a robot—an animated mechanical being.

Freedom and exceptional character are the two fundamental characteristics of the ludic experience that outline the contours of a symbolic space where rationality and seriousness are suspended, and where everything is possible—a dimension of extraordinary nature akin to a mythical time of origin, where nothing is stable and everything is still possible.[70] The recent experiments in performative games echo the unleashing of festive dimensions, allowing the performer to tap into the primordial sacred force of theater and the body by immersing in chaos, aiming to bring to a negative and destabilizing level what is foundational to theater, thus reinforcing and invigorating the sense of performative art, namely the relationship, the community, and the "everything is possible" ethos.

Returning to the performance *Robot*, we then arrive at the most chaotic scene, where dancers run across the stage while robots perform independent movements. This scene symbolizes the impossibility of total control and the emergence of chaos when human beings are involved. It represents the conflict and dynamic interaction between technological control and human spontaneity.

Thus, robots and humans enact both play and conflict, two fundamental anthropological categories. Play is understood not only in a ludic sense but as a crucial dimension of social and cultural life that facilitates power and social relations among individuals, and beyond. On the other hand, conflict (already evoked and analyzed in the previous paragraph) is studied as a key dimension of human interactions.[71]

Robots are metaphorical by nature, and it is important to leave the interpretation open; each artist/spectator is free to project whatever they wish onto these creatures. If one begins to feel something for a robot (pain, sadness, whatever it may be), then perhaps theater has achieved a goal that science has not yet fully attained.[72] The Nao robot used in this performance evokes the myth of Pinocchio:[73] both represent the being animated by the ambition to become "human", to acquire autonomy and sensitivity. Pinocchio, created by Carlo Collodi's imagination, has become a myth precisely because of its universal representation of the desire for humanity and the struggle between artificial nature and the aspiration for authenticity. The audience, witnessing the robot's falls and ambitions, experiences a range of emotional reactions, from disappointment to tenderness, reflecting the complex interaction between man and machine, between reality and aspiration, symbolized by Pinocchio.

The myth of Pinocchio is pertinent in this context not only for its quest for humanity but also for the broader reflection on technology and identity. In

Blanca Li's *Robot*, as in the story of Pinocchio, the tension between artificial creation and autonomous life is highlighted, a theme increasingly relevant in the era of advanced technology. Moreover, it offers a concrete and evident presence that cannot be diminished or ignored by the spectator's gaze, nor subjected to a phenomenological reduction or other epoché.[74] The technological virtuosity is not "exhibited" as in other robotic performances; instead:

> The dancers work eight hours a day. The robots, after one hour, go, 'Low battery. And then, bah. […]In the end, "ROBOT" is a celebration of the human body. "We can jump, we can turn, and our brain is constantly re-adjusting every movement so you don't fall," Ms. Li said. "No machine will ever be so amazingly rich in movement." And through the making of "ROBOT," she said: "I rediscovered dance. I realized how rich it is.[75]

It appears that the assertion of certain cognitive endowments, traditionally believed—rightly or wrongly—to be the exclusive domain of humans, finds its most challenging arena in the performing arts. While mechanical actors have demonstrated the capability to achieve an acceptable level of communication and expression in various theatrical and cinematic contexts, robots seem to encounter significant limitations in the realm of dance.[76] This limitation is evident when robots are restricted to imitating human movement, and as semi-humanoid partners, they are compelled to perform unnatural gestures if their "degrees of freedom" (DoF) are not adequately considered.

The DoF of a robot refer to the number of independent movements it can execute. In other words, they represent the number of independent variables necessary to fully determine the robot's position and orientation in space. Each degree of freedom corresponds to a possible direction of movement, which can be either a linear translation along an axis or a rotation around an axis.[77] The NAO model selected by the choreographer is undoubtedly a sophisticated and versatile mechanical agent, as it possesses 25 degrees of freedom distributed across different parts of its body, allowing for a wide range of complex movements.[78]

> Interaction is not merely an exchange of content; the "how," the "when," and the rhythm with which the synchronization of interventions occurs are of fundamental importance. It is akin to a harmonious dance of intentions and bodily expressions. Therefore, it is not only crucial that the robot performs the correct action, but it is also imperative that it does so in the correct manner and with the correct timing.[79]

The decline of theatrical mimesis[80] when observing a robot imitating human dance movements highlights the complex dynamics between technology, art, and human perception. Mimesis, or imitation, is fundamental in theater and performing arts, where an interpreter's ability to replicate reality or represent characters and situations is crucial for audience engagement. In dance, mimesis involves dancers expressing emotions, telling stories, and creating an illusion of life through body movements. When a robot is programmed to mimic human dance movements, it translates human actions into a machine-understandable language. Despite technological precision, several factors contribute to the decline of theatrical mimesis.

Lack of expressiveness is a key issue, as humans convey emotions through movements, including micro-expressions and subtle muscle tensions, which are challenging for robots to replicate perfectly. Even with accurate movement imitation, the absence of genuine expressiveness can make the robot appear mechanical and lifeless. The "uncanny valley" describes discomfort when an artificial entity seems almost, but not entirely, human.[81] Robots imitating human dance can fall into this valley, causing unease as the audience perceives the underlying artificiality. A human dancer's stage presence is tied to their physicality and liveliness. Authentic human movements derive from lived experiences and instant adaptability, which robots, regardless of sophistication, lack, resulting in performances that may feel soulless. Dance often involves deep interactions among dancers and between dancers and the audience, based on shared understanding of emotions and social cues. A robot can mimic movements but cannot fully engage in these human interactions, missing a fundamental performative dimension. While robots can technically imitate human dance movements with high precision, the decline of theatrical mimesis highlights the current technological limitations in replicating the authenticity, expressiveness, and emotional depth that characterize human performances.[82]

In the following paragraph, we will explore how, by facilitating an exchange of energy and paradoxically pursuing a mechanical poetics and aesthetics, the robot can become an adept artificial performer.

3.3 The Harmonious Dialogue between Man and Machine: The Pas de Deux of Huang Yi & Kuka

"The machined, or the evolved phase of artefacts, which attempts to imitate and behave as organisms by linking, coupling, morphing, merging and cross-fertilizations… the establishment of a rich ensemble of relations; networks, connections, breaks and unexpected links".[83]

In the context of dance performances involving industrial robots like the KUKA robotic arms, this quote from Deleuze and Guattari's *A Thousand Plateaus* vividly encapsulates the essence of integrating robotic elements into human artistic expression. One of the pioneering artists in this field is Pablo Ventura, who in 2002 presented the first chapter of his trilogy *De Humani Corporis Fabrica.*[84] This work was among the earliest to experiment with the integration of industrial robots into choreographic creations, specifically utilizing a KUKA robotic arm. Ventura seeks to convey a reflection on sentient beings through an audiovisual exploration and to illustrate potential "scenarios of the body" in the future. On one hand, new and unprecedented combinations of bodily mechanics, counterpoints, choreographic leitmotifs, and rhythms are achieved based on movement scores generated by the *Life Forms* software. On the other hand, innovative methods of interconnecting scenography, lights, videos, and sounds with non-human bodies that colonize the performance area are explored.[85]

Transcending an anthropocentric perspective that seeks to create robots in human likeness and attempts to teach machines the "correct" way to perform specific tasks, the dancer, choreographer, and engineer Huang-Yi poetically and harmoniously merges the art of dance with the science of mechanical engineering. In 2015, he conceptualized a duet with a robotic arm, which he programmed himself, entitled *Huang-Yi & Kuka*.[86]

The performance is structured through a series of choreographic sequences, each examining various facets of the interaction between human and machine. The introductory sequence features Huang-Yi entering the stage alone, executing slow and contemplative movements that evoke a sense of inquiry and discovery. This initial phase serves to acclimate the audience to the choreographic universe Huang-Yi has devised for his mechanical partner, distinguished by its precision and emotional sensitivity. A pivotal aspect of this creative paradigm shift is his approach to choreography: he constructs the movement sequences as if working with human dancers, while isolating the distinct components of the robot, thereby aligning with the mechanical capabilities of his robotic collaborator.

In the subsequent sequence, KUKA enters the stage, performing a series of movements that showcase its flexibility and precision. Following this parallelism and the presentation of these two brief solos, the audience is introduced to the encounter wherein Huang-Yi gradually approaches the robot, initiating a dance of exploration and mutual recognition. This sequence delves into the curiosity and marvel inherent in the meeting between human and machine, illustrating how two disparate entities can discover a common language through movement.

> The classical music (Bach, Mozart) used stands in captivating contrast to the modern buzzing and humming sounds KUKA makes. The stage remains quite dark throughout, like a dimly lit alternate universe. In that world, Huang Yi, who wears a dark suit, and KUKA, who is largely orange and arranged like a long multi-jointed arm, befriend each other and trade gestures. They use small lights to illuminate each other and the space. In the timing and shape of KUKA's movements, one senses hesitation and shyness as the two are introduced. One further imagines exchanged glances, curiosity, skepticism, and ultimately friendship.[87]

In the subsequent sequence, titled "Silent Dialogue" Huang Yi and KUKA initiate a series of synchronized movements reminiscent of a silent conversation. Huang Yi's gestures are fluid and delicate, while KUKA's movements are precise and methodical. Both find harmony through "listening" and sharing the same performance space. This section symbolizes the paradoxical possibility of communication and understanding between different entities, suggesting a dialogue that transcends words—a domain where dance excels artistically—and an understanding linked to the awareness of the other's presence, despite the absence of mutual gaze.

In dance, listening to the other refers to the dancers' ability to perceive and respond to their partners' movements, energy, and intentions. This type of listening is not solely auditory but also corporeal and spatial, requiring a high level of awareness and sensitivity. This is why the Taiwanese choreographer chooses Contact Improvisation as the style for his duet.

Nancy Stark Smith, one of the founders of Contact Improvisation, describes the concept of "kinesthetic dialogue" as a fluid and continuous interaction between dancers who communicate through movement and physical contact. In this dialogue, each dancer is keenly aware of their partner's intentions, movements, and energy, creating a non-verbal communication flow fundamental to this improvisational dance form. Smith described this experience as a moment when "every part of me feels very alive—lit up, awake, present, ready, both at ease and on the edge, honed and softened, deepened and extended, challenged and soothed, physically and energetically attuned to myself and to others".[88]

The bodily energy of choreographic gestures in contemporary dance, particularly in Contact Improvisation, plays a crucial role in expression and communication. Rudolf Laban, in his movement analysis system, identified several factors influencing the energy of movements, such as weight, space, time, and flow. "Kuka, his crane-like companion, has joints of steel instead of flesh, but this German-made robot, painstakingly programmed by Mr. Huang—it takes him 10 hours to produce one minute of movement—is oddly poetic, at least initially".[89]

These elements determine the energetic quality of a gesture and how it can convey emotions and narratives.[90] The conscious management of energy allows dancers to modulate the intensity of their performances, creating contrasts and dynamics that capture the audience's attention.[91] The practice of this type of dance, which involves constant awareness of other dancers and the surrounding space, can be seen as a choreography of identification or empathic choreography, where dancers continually adapt to the movements, direction, speed, and energy of others.[92]

It is now recognized that the notion of 'object' seems inadequate to describe contemporary technology, which lacks a rigid identity or a univocal form and is capable of continuous transformation. As previously stated, the robotic agent fits Bell's definition of a performative object.[93] The term 'machine' is also reductive when referring to a robot. Human experiential and artistic learning has indeed changed, shifting from being linked to the individual's mind and body before the mass introduction of technological means, to now being tied to the use of devices as a means of accessing sensory and physical knowledge.

This transformation can be better understood through postmodern theories. For example, Bruno Latour, with his *Actor-Network Theory* (ANT), emphasizes that technological objects are not mere tools but actors that actively participate in the construction of social networks. Latour argues that technologies and humans form a network of relationships in which both actors reciprocally influence their existence and development. Technological objects, therefore, are seen as dynamic entities that co-create the social world alongside humans.[94]

Similarly, Donna Haraway in her *Cyborg Manifesto* argues that new technologies are dissolving the boundaries between human and machine, nature and culture, leading to a new form of hybrid existence.[95] Human 'knowledge' thus evolves along with technology, transforming into an interactive process continuously renegotiated between the individual and the device.

This is not the usual ballet where the robot mimics human gestures or vice versa. Instead, we are witnessing an attempt at synergistic hybrid creation.

> This is an example of co-construction of meaning between the human and the non-human, which Donna Haraway describes as sympoiesis in her book *Staying with the Trouble*.[96], Sympoiesis, a simple word meaning 'making with', implies that nothing happens alone; nothing is truly autopoietic or self-organized. Sympoiesis refers to complex, dynamic, responsive, situated, and historical systems. It encompasses autopoiesis, unfolding and extending it in a generative way.[97]

Returning to the performance, we arrive at the "Dance of Symbiosis", where the two performers execute interdependent movements, with each gesture influencing and completing the other. This sequence is characterized by an alternation of moments of harmony and tension, representing the evolution of the symbiotic relationship between contemporary humans and machines.

> Mutualism, as a form of symbiosis, refers to interactions between different species that are mutually beneficial. These interactions play a crucial role in ecological and evolutionary processes, fostering cooperation that enhances survival and adaptation among the species involved. The concept is fundamental to understanding the complex interdependencies that characterize natural ecosystems.[98]

Toward the end, in the segment titled "Industrial Rhythm", the movements become faster and more rhythmic, with Huang Yi and KUKA executing a complex choreography reminiscent of industrial machine operations. This segment highlights the strength and precision of technology while also showcasing the human ability to adapt to mechanical rhythms and processes.

In the physical world, a robot must possess sensory, kinesthetic, and sensorimotor capabilities, no matter how specific or elementary. It must dynamically perceive the object it is working on and the environment it moves within. In this way, it learns from its actions and thus has a body with specific attributes, a motor system, feedback, and coordination—and in some cases, a body of knowledge and the ability to learn—much like a dancer.

The performance culminates in an "Emotional Epiphany", characterized by slow and deeply emotional movements, with Huang Yi and KUKA seemingly merging into a single entity. The choreography is delicate and touching, culminating in a symbolic embrace between human and machine, representing the peak of their connection. Physical contact is replaced by the dancer's surrender to the support of KUKA. The human relies on his mechanical partner, whom he has personally 'trained'. Although he is aware that he is not being seen by the robot, he is certain of being perceived, located in space, and he trusts by surrendering the weight of his body to the robotic arm.

> Some lonely children create invisible friends. As a solitary boy growing up in Taiwan, the choreographer Huang Yi pined for a robot. He had good reason: In his poignant artist's statement for his show "Huang Yi & Kuka," at 3LD Art & Technology Center, he reveals that when he was 10, his wealthy parents became bankrupt and suicidal. The family moved to a 40-square-foot room; Mr. Huang, feeling compelled to be a perfect child, masked his emotions to become the picture of obedience. In other words, a robot. [...] In "Huang Yi & Kuka," that childhood wish comes true.[99]

This performance not only merges modern dance and visual arts with the field of robotics, conceptualizing the fascination of the union between man and machine, but it can also be classified under the genre of puppet theater known as "object theater".[100] This is a type of performative practice that seeks to communicate through the metaphorical value, evocative power, and poetic force of objects. When we refer to "objects", we mean items that have been or are part of our daily lives, with all the imaginative and, in the case of robots, dystopian connotations they contain (Figure 3.3).

Objects are like documents extracted from reality, and the artist interrogates them as witnesses of our situated and constructed collective representations. The expressions "real object" and "true object" are part of a discourse of definition and legitimation specific to object theater. First, looking at its definition: unlike puppet theater, which represents a "similar brother" object

Figure 3.3 Huang Yi & KUKA. Photo by Jacob Blickenstaff. Courtesy of Huang Yi Studio + (Taiwan).

theater is based on the manipulation of objects that were not designed for the theater but are used in the theater. Second, it was legitimized because the term "object theater" appeared in the early 1980s, mainly by artists trained in the plastic arts, who opposed the invasion of objects in consumer society, thus creating this "genre" as an "act of resistance against all forms of obsolescence and a touching desire to say something about human fragility on stage":[101]

The idea of a puppet is erratic, inconsistent, and changes according to the places and times, adapting to the cultures in which it is inserted. Defining a robot as a sophisticated puppet is not a downgrade, considering that as a performative partner, although effective as in this choreography, it lacks self-awareness and the autonomy to replicate the gesture of its partner.

> Thus, our worldview begins with the perception received through sensors but is processed based on culture, which also includes the emotional archive of our species filtered through the individual, forming their personal weltanschauung. It is not merely an archive of experiences, even if classified by taxonomy according to normative reasoning criteria, as a synthetic creature might do.[102]

The puppet, therefore, serves as a symbol of the "illusion of life", an almost esoteric object that sometimes seems to embody an entity. What emerges and functions in this performance is the relationship between the dancer-puppeteer (the one with a gaze, a point of view), the performing object (rather than manipulated), and the spectator (the one who watches).[103]

This ongoing process of definition, symbolically translated in this dance, can also be analyzed from an anthropological perspective, where technology is not simply a set of tools but an integral part of our cultural evolution. In anthropology, the concept of "techno-sociality" describes how technology influences social relationships and the construction of identity. According to Sherry Turkle, digital technologies profoundly shape our interactions and sense of self, creating new forms of connection and isolation.[104] Additionally, Tim Ingold highlights that technology is not seen as an external element to human culture but as something that co-evolves with it, continuously transforming our ways of learning and interacting with the world.[105]

This evolution in learning reflects a broader transformation in cultural and social practices, where the integration of technological devices has become essential for mediating human experiences. Thus, technology not only amplifies the cognitive and physical capacities of the individual but also redefines their aesthetic taste and conception of art.

The examples presented so far lead us toward artistic paths that aim to dissolve the physical archetypes in which the human is enclosed and to nullify the dualism that opposes and elevates humans above what is non-human. Through theatrical encounters with the other, we move away from the egoic (self-referential) and sometimes entropic human condition. This process can be understood through Maurice Merleau-Ponty's concept of "intercorporeality", which emphasizes how the body is a meeting point between self and other, and how mutual understanding is created through bodily interaction.[106]

One might wonder if the artist's impulse to create with these new entities has complex reasons leading to a rethinking of what can be defined as a simple object and what can be defined as a non-living—or biological—entity endowed with its own "singularity". Philosopher Giorgio Agamben, in his work on aesthetics and politics, explores how new technologies can redefine the concepts of life and singularity, suggesting that such entities may have a form of life that challenges traditional ontological categorizations.[107]

Additionally, performance studies provide theoretical tools to analyze these dynamics. Peggy Phelan, for instance, speaks of performance as a mode that exists in the moment, challenging notions of reproducibility and permanence.[108] This concept can be extended to interactions with technological entities, where performance becomes a site of encounter and transformation rather than a mere act of representation.

Finally, it is noteworthy how the approach of an Eastern artist results in an artistic outcome distinctly different from those mentioned earlier by European artists. In this performance, the influence of animism is evident—the belief that inanimate objects may possess a soul or consciousness, which can be applied to robots in the contemporary era. This conception has deep roots in traditional cultures where animism is an integral part of the relationship between humans and the surrounding world.[109] In the field of robotics, animism manifests in the way robots are designed and perceived as entities endowed with agency and intentionality. Huang Yi seems to have developed an emotional bond with Kuka, seeing it as a companion rather than a mere machine.[110]

> On stage, the robot enchants us with its seductive power or frightens us with its call to a dystopian future, but it does not leave us indifferent. It sees but does not watch, listens but does not always understand, perceives but does not always recognize: it lacks consciousness or experience, and for now, its power is based on memory and infallible logic.[111]

Notes

1 See L. Damiano, P. Dowell, *Living with Robots: An Essay on Artificial Empathy*, Harvard University Press, Cambridge, MA, 2017, pp. 10–12.

2 For further insight into the Eastern conception of the robot, see: Y. Sone, *Japanese Robot Culture*, Palgrave Macmillan, New York, 2016 and I.H. Tuan, *Robot Theatre and AI Films. In: Pop with Gods, Shakespeare, and AI.* Palgrave Macmillan, Singapore, 2020, doi: 10.1007/978–981-15-7297-5_9.

3 The latter, in particular, won the Olivier Award in 1989 and 1990 for Best New Musical and enjoyed renewed popularity in the UK with a revival in 2015.

4 We deliberately use the adjective experimental "to account for a multiplicity of processes that often lack common traits; not, for instance, the use of measuring instruments in their specific relationship with a theory, nor for the objectives for which the experiments are conducted. The term experimental appears to contrast these processes with observations on one hand and purely theoretical reflection on the other", V. Schiaffonati, *Computer, Robot, and Experiments*, Meltemi, Milano, 2020.

5 After this brief terminological preamble, we will address several examples that primarily concern the use of robotic agents, intentionally leaving aside, not due to a lack of importance but because of the extensive existing literature on the subject, the renowned Human Theater Project by Oriza Hirata. For further information on this topic, see Z. Paré, *Des robots acteurs* in Transhumanité, L'Harmattan, Paris, 2013.

6 T. Riccio, Sophia Robot: An Emergent Ethnography, in *TDR/The Drama Review*, 65(3), 51–52, Sptember 2021, doi: 10.1017/S1054204321000319.

7 Especially when this is hardly distinguishable from a human me, some well-known actroids come to mind, such as Geminoid F, or Thomas Melle's double used in the Uncanny Valley play chosen as a case study or, again, the performer *Sophia Robot: An Emergent Ethnography*, created by Hanson Robotics, see T. Riccio, Sophia Robot: An Emergent Ethnography, in *TDR/The Drama Review*, 65(3), pp. 51–52, Sptember 2021, doi: 10.1017/S1054204321000319.

8 A. Mayor, *Gods and Robots: Myths, Machines, and Ancient Dreams of Technology,* Princeton University Press, Princeton, NJ, 2020.

9 L.A. Erscoi, A. Kleinherenbrink, O. Guest, Pygmalion Displacement: When Humanising AI Dehumanises Women, in *Science Letter*, 522, 17 November 2023. Gale Academic OneFile, link.gale.com/apps/doc/A772499910/AONE?u=anon~e4dd5017&sid=sitemap&xid=9ab199e7, 2023, [accessed 4 June 2024].

10 J. Alter, *Sociosemiotic Theory of Theater*, University of Pennsylvania Press, Philadelphia, 1990.
11 The trio, based at Berlin's Hebbel am Ufer theatre, has long developed a form of documentary theatre that pokes fun at the perception and structural integrity of reality. https://www.rimini-protokoll.de/website/en.
12 H. Jenkins, *Convergence Culture: Where Old and New Media Collide*, University Press, New York, 2006, pp. 140–144.
13 See J.D. Bolter, R. Grusin, *Remediation: Understanding New Media*, MIT Press, Cambridge, MA, 2000.
14 P. Eckersall, Towards a Dramaturgy of Robots and Object-Figures, in *TDR/The Drama Review*, 59(3 (227)), 123–131, 2015, doi: 10.1162/DRAM_a_004742015.
15 E. Fuoco, Could a Robot Become a Successful Actor? The Case of Geminoid F, in *Acta Universitatis Lodziensis Folia Litteraria Polonica*, 65(2), 203–219, 2022, doi: 10.18778/1505–9057.65.11.
16 Reference is made to Western society, which with a considerable time lag is still in the process of integration of robots, despite Eastern cultural society, which has always seen them as characters of the childhood imagination, but also as a resource for the future of humanity. See S. Dixon, *Digital Performance: A History of New Media in Theater, Dance, Performance Art, and Installation*, The MIT Press, Cambridge, MA, 2007, p.10.
17 See J. Parker-Starbuck, *Introduction: Why Cyborg Theatre. In Cyborg Theatre*, Palgrave Macmillan, London, 2011.
18 J. Bell, *Puppets, Masks, and* Performing Objects, MIT Press, Cambridge, MA, 2001, p. 15.
19 Concept, Text & Direction: Stefan Kaegi. Text/Body/Voice: Thomas Melle. Equipment: Evi Bauer. Animatronic: Chiscreatures Filmeffects GmbH, https://www.rimini-protokoll.de/website/en/project/unheimliches-tal-uncanny-valley.
20 This new show by Stefan Kaegi is part of the trend in stage productions of the past ten years that place the humanoid robot at the heart of live performances. Let us mention only the most accomplished productions, such as *Spillikin, a Love Story* by Pipeline Theatre (2015) or the shows by Oriza Hirata, *Sayonara Ver. 2*(2012), *Three Sisters Android Version* (2012), and *Metamorphoses. Android Version* (2014). These robots, with their strong human resemblance, are characterized by highly developed technological sophistication". I. Pluta, *Quand les frontières entre l'original et la copie se brouillent. La Vallée de l'étrange de Stefan Kaegi*, in Critiques. Regard sur la technologie dans le spectacle 2020 vivant. Carnet en ligne de Theatre in Progress URL: http://theatreinprogress.ch/?p=638#more-638.
21 C. Faletti, G. Sofia, V. Jacono, *Theatre and Cognitive Neuroscience*, Bloomsbury Publishing, London, 2016.
22 The Uncanny Valley phenomenon, introduced by Ernst Jentsch and further theorized by Masahiro Mori, describes the discomfort people feel when interacting with objects that appear almost human but lack the empathic familiarity of real humans. This discomfort turns to fear or revulsion when artificial beings resemble humans closely without being fully lifelike. However, when these artifacts become indistinguishable from living beings, the level of acceptance rises, potentially allowing for emotional interaction. See E. Fuoco, Could a Robot Become a Successful Actor? The Case of Geminoid F, in *Acta Universitatis Lodziensis Folia Litteraria Polonica*, 65(2), 203–219, 2022, p. 213.
23 See I. Pluta, Robot-Author. Composite Dramaturgy in Stefan Kaegi's Uncanny Valley, in G. Guccini, C. Longhi and D. Vianello (eds), *Creating for the Stage and Other Spaces: Questioning Practices and Theories, Arti della Performance: orizzonti e culture*, Alma DL – University of Bologna Digital Library, n. 13, 2021 – ISBN 97888549707172021.

24 Geminoid from the Latin *geminus*, meaning twin or double and adding 'oides' which indicates "similarity" or being a twin. "As the name suggests, a Geminoid is a robot that functions as a duplicate of an existing person. It appears and behaves as a person and is connected to that person by a computer network. Geminoids extend the applicable field of android science. Androids are designed for studying human nature in general. With Geminoids, we can study such personal aspects as presence or personality traits, tracing their origins and implementation into robots", S. Nishio, H. Ishiguro, N. Hagita, Can a Teleoperated Android Represent Personal Presence? A Case Study With Children, in *Psychologia*, 50(4), 330–343, 2007, p. 5.

25 N. K. Hayles, How We Became Posthuman: Virtual Bodies in Cybernetics, Literature, and Informatics, The University of Chicago Press, Chicago - London 1999.

26 It was decided to place at the center of this brief reflection the well-known Rimini Protokoll's *Uncanny Valley* play, already much analyzed in the performance studies field, instead of Oriza Hirata's plays, because, unlike the latter, the German collective's performance—or better the performative show—was staged using a single non-human actor. There is no interaction between human and artificial actors, or between humanoid robots. There is a robot that plays the part of a human, looks like a human and addresses the audience via a monologue, along with its double simultaneously playing on video. A sort of theatre on the edge, devoid of the human, which paradoxically enhances the importance of its presence, in this case embodied by the audience, see J. Parker Starbuck, *Cyborg Theatre : Corporeal/Technological Intersections in Multimedia Performance*, Palgrave Macmillan, New York, 2011, p. 141.

27 M. Corvin, *L'homme en trop. L'abhumanisme dans le théatre contemporain*, Les Solitaires Intempestifs, Besançon, 2014, p. 115.

28 See A. Caronia, *Il CYBORG. Saggio sull'uomo artificiale,* ShaKe editore, Milano, 2001.

29 V. Turner, *From Ritual to Theatre: The Human Seriousness of Play*, PAJ, Cambridge, MA, 1982, p. 35.

30 K. Barad, *Meeting the Universe Halfway: Quantum Physics and the Entanglement of Matter and Meaning*, Duke University Press, Durham, NC, 2007, p. 206.

31 K. Barad, Posthumanist Performativity: Toward an Understanding of How Matter Comes to Matter, in *Signs: Journal of Women in Culture and Society*, 28(3), 801–831, 2003. doi: 10.1086/345321, [accessed 4 June 2024].

32 K. Barad, *Meeting the Universe Halfway: Quantum Physics and the Entanglement of Matter and Meaning*, Duke University Press, Durham, NC, 2007, p. 245.

33 K. Barad, *Meeting the Universe Halfway: Quantum Physics and the Entanglement of Matter and Meaning*, Duke University Press, Durham, NC, 2007, p. 210.

34 R. Schechner, 6 Axioms for Environmental Theatre, in *The Drama Review: TDR*, 12(3), 41–64, 1968, *JSTOR*, doi: 10.2307/1144353, [accessed 4 June 2024], p. 41.

35 The word *théatron*, from which the modern word "theater" derives, first appears in 4th-century B.C. Greek literary texts, stemming from the verb *theàomai*, "to see", which designated both the places where one attended a spectacle and the crowd of spectators watching the spectacle.

36 See B.H. Bratton, *The Revenge of the Real: Politics for a Post-Pandemic World*, Duke University Press, Durham, NC, 2021.

37 J. Collins, J. Jervis, Document: 'On the Psychology of the Uncanny' (1906): Ernst Jentsch, in J. Collins, J. Jervis (eds), *Uncanny Modernity*. Palgrave Macmillan, London, 2008, doi: 10.1057/9780230582828_12, p. 15.

38 M. Heidegger, *The Question Concerning Technology and Other Essays*, Garland Publishing, Inc., New York and London, 1977, p. 12.

39 M. Heidegger, *The Question Concerning Technology and Other Essays*, Garland Publishing, Inc., New York and London, 1977, p. 12.
40 M. Heidegger, *The Question Concerning Technology and Other Essays*, Garland Publishing, Inc., New York and London, 1977, p. 17.
41 M. Heidegger, *The Question Concerning Technology and Other Essays*, Garland Publishing, Inc., New York and London, 1977, p. 22.
42 M. Heidegger, *The Question Concerning Technology and Other Essays*, Garland Publishing, Inc., New York and London, 1977, p. 28.
43 R. Wolin, Being and Time, in *Encyclopedia Britannica*, 22nd April 2024, https://www.britannica.com/topic/Being-and-Time.
44 J.W. Carroll, N. Markosian, *An Introduction to Metaphysics*, Cambridge University Press, 2010, p. 45.
45 M. Heidegger, *The Question Concerning Technology and Other Essays*, Garland Publishing, Inc., New York and London, 1977 p. 35.
46 See https://classics.mit.edu/Aristotle/poetics.1.1.html.
47 P. Demetz, *Brecht; A Collection of Critical Essays*, Prentice-Hall, New Jersey, 1962.
48 In the realm of performance art, glitches can transform unintended malfunctions into deliberate artistic expressions, creating a unique and compelling aesthetic. This approach not only adds a layer of unpredictability and complexity to the performance but also invites the audience to engage with the material in new and reflective ways. For example, artists like Rosa Menkman have explored the potential of glitch art to reveal the underlying structures and processes of digital media, turning imperfections into visually striking and thought-provoking pieces, see R. Menkman, *The Glitch Moment(um)*, Institute of Network Cultures, Amsterdam, 2011.
49 O. Goriunova, A. Shulgin, *Glitch, in Software Studies: A Lexicon*, in Matthew Fuller (ed.), MIT Press, Cambridge, MA, 2008.
50 See G. Cox, Postscript on the Post-digital and the Problem of Temporality, in D.M. Berry, M. Dieter (eds), *Postdigital Aesthetics*, Palgrave Macmillan, London, 2015. doi: 10.1057/9781137437204_12.
51 Through deliberate actions and context, these objects become intentional signs, communicating complex ideas and narratives beyond their mere material form. This concept is significant in understanding how symbols and performances convey deeper cultural and human experiences.
52 See D. Plassard, *L'automate et la marionnette: Histoire et analyse d'un rêve*, L'Harmattan, Paris, 1992.
53 See D. Plassard, *L'automate et la marionnette: Histoire et analyse d'un rêve*, L'Harmattan, Paris, 1992.
54 See P. Reilly, *Automata and Mimesis on the Stage of Theatre History*, Palgrave Macmillan, New York, 2011.
55 See H. von Kleist, *Über das Marionettentheater*, Reclam, Stuttgart, 1996.
56 See E. Kac, *Telepresence and Bio Art: Networking Humans, Rabbits, and Robots*, University of Michigan Press, Ann Arbor, 1997.
57 C. Breazeal, R.A. Brooks, J. Gray et al., Robots in the Wild: Observing Human-Robot Social Interaction Outside the Lab, in *AI Magazine*, 25(4), 33–42, 2003.
58 See A. Elliott, *The Culture of AI. Everyday Life and the Digital Revolution*, Routledge, Abingdon-on-Thames, 2019.
59 Z. Paré, *L'empathie articielle,* Interview followed by a discussion moderated by Isabelle Cossin, Ioana Ocnarescu, Estelle Berger, and Frédérique Pain, with the participation of Alexandre Mazel (SoftBank Robotics) and Dominique Dauff (Orange). Commentary: Cultural gap of the speakers locked into a rehashed and obsolete discourse on social robotics, followed by exercises in contortion to try

to adhere to more realistic and less mystifying common-sense notions regarding service robotics!, p. 24. See https://www.zavenpare.com/.

60 This capacity for empathetic engagement with robots is a widely discussed topic in contemporary literature. For example, in See L. Damiano, P. Dowell, *Living with Robots: An Essay on Artificial Empathy*, Harvard University Press, Cambridge, MA, 2017, the authors explore how robots can induce empathetic reactions in humans. Similarly, Anna Donise, *Critica della Ragione Empatica: Fenomenologia dell'Altruismo e della Crudelt*, Il Mulino, Bologna, 2020 analyzes the mechanisms through which empathy can be evoked, even towards inanimate objects. These texts, along with other studies on artificial intelligence provide a comprehensive overview of the empathetic dynamics between humans and robots, emphasizing the importance of understanding how interactions with non-living agents can influence our emotional and behavioral responses.

61 https://www.blancali.com/spectacle/robot-film-directors-cut/.

62 Blanca Lì examined many models before settling on NAO, a companion robot manufactured by the French company Aldebaran. "They are the size of a one-year-old child, and I found them very cute," she said. "They talk, they can recognize you. They have a beautiful kind of movement, even though they are robots and have many limitations". In G.G. Kourlas, 'ROBOT', Blanca Li Uses Machines to Celebrate the Body, in *The New York Times,* Tuesday, 9th June 2015, https://www.nytimes.com/2015/06/09/arts/dance/in-robot-blanca-li-uses-machines-to-celebrate-the-body.html.

63 The eight dancers were accompanied by a ten-piece orchestra of unique sculptural as well as musical gadgetry. Imagine a sculpture somehow escaping the confines of the Museum of Modern Art and then appearing on a NYC subway platform playing a violin and there you have it. These gadgetries came in a multitude of styles, shape and color. There was a yellow and green boxy thing that played the drums as well a metallic humanesque object that played the violin and these were just two. The music and these imaginative creations were the joint product of the Japanese design team Maya Denki and composer Tao Gutierrez.

64 R. Notte, *You Robot. Antropologia della vita artificiale*, Vallecchi, Firenze, 2005.

65 A. Guillot *Danse avec les robots* in Philosophie Magazine, 5th January 2021, https://www.philomag.com/articles/danse-avec-les-robots.

66 He identifies four phases of the simulacrum: The representation of a basic reality. The perversion of a basic reality. The masking of the absence of a basic reality. The simulacrum, which has no relation to any basic reality, is its own pure simulation. See J. Baudrillard, *Simulacres et Simulation*, Galilée, Paris, 1981.

67 The technicians are always on hand, and nothing is hidden from the audience. For Blanca Li, exposing the show's technical difficulties mirrors modern life, where smartphones break, and Wi-Fi is never as reliable as you need it to be.

68 See J. Huizinga, Homo Ludens: A STudy of the Play, *Element in Culture*, Beacon Press, New York 1971.

69 Ibidem.

70 See R. Caillois, *Les jeux et les hommes. La masque et la vertige*, Gallimard, Paris, 1967.

71 A.W. Wolfe, H. Yang, *Anthropological Contributions to Conflict Resolution*, University of Georgia Press, Athens, 1996.

72 Consider the increasing attention being dedicated to the ethics of form and the affordance of robots. See M. Coeckelbergh, *Robot Ethics*, The MIT Press, Cambridge, MA, 2022.

73 A. Di Martino, *Pinocchio. Figura del mito*, Mimesis, Milano, 2023.

74 J.-P. Sartre, *L'Être et le Néant: Essai d'ontologie phénoménologique*, Gallimard, Paris, 1943.

75 G. Kourlas, *'ROBOT,' Blanca Li Uses Machines to Celebrate the Body*, The New York Times, 8th June 2015, https://www.nytimes.com/2015/06/09/arts/dance/in-robot-blanca-li-uses-machines-to-celebrate-the-body.html.

76 See E. Fuoco, Could a Robot Become a Successful Actor? The Case of Geminoid F, in *Acta Universitatis Lodziensis Folia Litteraria Polonica*, 65(2), 203–219, 2022.
77 J. Craig, *Introduction to Robotics: Mechanics and Control*, Pearson, Prentice Hall, NJ, 2005.
78 NAO's head has two degrees of freedom, allowing it to move along two main axes: tilt (up and down) and rotation (left and right). This enables it to look around, enhancing its ability to perceive and interact with the environment and people. NAO's arms are equipped with five degrees of freedom each, for a total of 10 degrees of freedom. Each arm can move the shoulder on two axes (rotation and tilt), the elbow on one axis (flexion), and the wrist on two axes (rotation and tilt). This configuration allows NAO to perform complex movements, such as picking up objects, gesturing, and carrying out actions that require precision. Each of NAO's hands has one degree of freedom, allowing it to open and close its fingers. Although this movement is relatively simple, it is essential for gripping and manipulating objects. NAO's legs, each with five degrees of freedom, are designed to enable it to walk, run, and maintain balance. Each leg has two degrees of freedom at the hip (rotation and tilt), one degree of freedom at the knee (flexion), and two degrees of freedom at the ankle (rotation and tilt). This joint complexity allows NAO to perform stable and coordinated movements, making it capable of navigating different surfaces and adapting to various terrains. See https://unitedrobotics.group/en/robots/nao?utm_source=aldebaran&utm_medium=referral and D. Gouaillier, V. Hugel et al., Mechatronic Design of NAO Humanoid, in *IEEE International Conference on Robotics and Automation*, IEEE, 2009, pp. 769–774.
79 E. Fuoco, Could a Robot Become a Successful Actor? The Case of Geminoid F, in *Acta Universitatis Lodziensis Folia Litteraria Polonica*, 65(2), 203–219, 2022.
80 Aristotle, *Poetics*, Dover Publications, Mineola, NY, 1996.
81 See M. Mori, Bukimi no Tani [The Uncanny Valley], in *Energy*, 7(4), 33–35, 1970.
82 See G. Riva, C. Galimberti, *Towards Cyberpsychology: Mind, Cognitions and Society in the Internet Age*, IOS Press, 2001.
83 G. Deleuze, F. Guattari, *A Thousand Plateaus: Capitalism and Schizophrenia*, University of Minnesota Press, Minneapolis, 1987.
84 https://www.ventura-dance.com/works/de-humani.
85 Kubic's Cube Performance/Installation epilogue to the trilogy De Humani Corporis Fabrica (2002–2006).

The Installation (3–4 hours): the robot "Kubic" hangs in a dark space surrounded by sounds of a jungle. The public moves freely in space and regards the robot from different perspectives. A motion tracking software registers the whereabouts of the public and triggers the robot to react to the approach and nearness of the spectators. An ambivalence takes place between the sounds of organic life in a jungle and a machine and the public's perception of "Kubic" fluctuates between a kinetic sculpture or a living organism. The performance (20 minutes.) The choreographer Pablo Ventura operates the robot 'live' to industrial sounds especially composed for the piece by Francisco López, creating differentiated mechanistic atmospheres that invite the viewer to discover Kubic's Cube from different perspectives, https://www.ventura-dance.com/works/kubic.
86 https://www.sozoartists.com/kuka.
87 C. Morgan, *One tool dancing with another: Huang Yi & KUKA*, 1st October 2015, https://criticaldance.org/one-tool-dancing-with-another-huang-yi-kuka/.
88 D. Koteen, N. Stark Smith, *Caught Falling: The Confluence of Contact Improvisation, Nancy Stark Smith, and Other Moving Ideas*, Contact Editions,Toronto, 2008.
89 *Review: 'Huang Yi & Kuka,' a Pas de Deux Between Human and Machine*, https://www.nytimes.com/2015/02/13/arts/dance/review-huang-yi-amp-kuka-a-pas-de-deux-between-human-and-machine.html.

90 R. Laban, *Laban's Principles of Dance and Movement Notation*, Macdonald & Evans, Estover, Plymouth, 1947.
91 See J. Hodgson, *Mastering Movement: The Life and Work of Rudolf Laban*, Routledge, 2001.
92 C.J. Novack, *Sharing the Dance: Contact Improvisation and American Culture*, University of Wisconsin Press, Madison, 1990.
93 J. Bell (ed.), *Puppets, Masks, and Performing Objects*, The MIT Press, Cambridge, MA, 2001, p. 20.
94 See B. Latour, *Reassembling the Social: An Introduction to Actor-Network-Theory*, Oxford University Press, Oxford, UK, 2005.
95 D. Haraway, A Cyborg Manifesto: Science, Technology, and Socialist-Feminism in the Late Twentieth Century, in *Simians, Cyborgs, and Women: The Reinvention of Nature*, Routledge, New York, NY, 1995.
96 See D. Haraway, *Staying with the Trouble: Making Kin in the Chthulucene*, Duke University Press, Durham, NC, 2016.
97 See E. Fuoco, *Né qui, né ora: peripezie mediali della performance contemporanea*, Ledizioni, Milano, 2022.
98 A.E. Douglas, Th*e Symbiotic Habit*, Princeton University Press, Princeton, NJ, 2021.
99 G. Kourlas, *Review: 'Huang Yi & Kuka', a Pax de Deux Between Human and Machine*, in The New York Times, 12th February 2015, https://www.nytimes.com/2015/02/13/arts/dance/review-huang-yi-amp-kuka-a-pas-de-deux-between-human-and-machine.html.
100 C. Corniquet, M. Rhéty, *L'objet vrai: Précis des objets dans le théâtre d'objets d'Agnès Limbos*, à partir de Troubles in Agôn, Revue des arts de la scene, 15 April 2020, doi: 10.4000/agon.2077.
101 C. Corniquet, M. Rhéty, *L'objet vrai: Précis des objets dans le théâtre d'objets d'Agnès Limbos*, à partir de Troubles in Agôn, Revue des arts de la scene, 15 April 2020, doi: 10.4000/agon.2077.
102 P. Dario, G. Anerdi, *Compagni di viaggio. Robot, androidi e altre intelligenze*, Codice, Torino, 2022, p. 167.
103 See C. Corniquet, M. Rhéty, *L'objet vrai: Précis des objets dans le théâtre d'objets d'Agnès Limbos*, à partir de Troubles in Agôn, Revue des arts de la scene, 15 April 2020, doi: 10.4000/agon.2077.
104 See S. Turkle, *Alone Together: Why We Expect More from Technology and Less from Each Other*, Basic Books, New York, 2011.
105 See T. Ingold, *The Perception of the Environment: Essays on Livelihood, Dwelling and Skill*, Routledge, 2000.
106 M. Merleau-Ponty, *Phénoménologie de la perception*, Gallimard, Paris, 1945.
107 G. Agamben, What is a Destituent Power, in *Environment and Planning D: Society and Space*, 32, 65–74, 2014.
108 See P. Phelan, *Unmarked: The Politics of Performance*, Routledge, New York, 1993.
109 N. Bird-David, Animism Revisited: Personhood, Environment, and Relational Epistemology, in *Current Anthropology*, 40(S1), S67–S91, 1999.
110 See S. Turkle, *Alone Together: Why We Expect More from Technology and Less from Each Other*, Basic Books, New York, 2011.
111 See E. Fuoco, *Né qui, né ora: peripezie mediali della performance contemporanea*, Ledizioni, Milano, 2022. p. 114.

4 Rethinking Linear Perspective in Contemporary Theatrical Studies

This chapter delves into the heuristic effectiveness of theatricality as a paradigm for interpreting social reality as a complex and conflictual relational field. It explores how performance, studied in theater, magic, and ritual, becomes an anthropological subject. The representation of deviation in media shapes public perceptions and influences cultural narratives. It discusses the role of deviation in challenging societal norms, paralleling the use of technological devices in theatrical innovation. It highlights how digital platforms enable greater audience participation, fostering a dynamic and interactive community. This shift from traditional to innovative practices redefines societal norms through technological integration and artistic exploration, emphasizing the transformative potential of performing arts.

4.1 Performative Deviations: Exploring the Intersection of Technology and Art

It appears to be proven, following the examples brought to attention, the heuristic effectiveness of theatricality understood as a paradigm for interpreting social reality as a complex and conflictual relational field, an area in which subjects are engaged in a continuous redefinition of themselves and their relationships, reconsidering also the general terms of the reality in which they live. It is precisely for this reason that performance has become the object of anthropological sciences, which have investigated it in the theater and in related fields such as magic, ritual, and other pro-theatrical and para-theatrical forms. These dimensions all possess characteristics of exceptionality, collective and communal dimensions, and in some ways a symbolic manipulation of the conception of the human and reality. Many refer to the power of theater as magic, a magic without deceit, a field of possibility that unfolds and opens a scenario in an unreal place and time.[1]

Certainly, we have observed how even the perception invoked in the hybrid theatrical experiences reported is paradoxical, a perceptual fantasy that requires the real but only insofar as it is incomplete or unposed.[2] The representation of deviation in media and popular culture can shape public perceptions

DOI: 10.4324/9781032676937-5

and influence cultural narratives. In summary, deviation is a key concept for understanding the evolutionary process that the performing arts are undergoing. Deviation, in sociology and anthropology, refers to behaviors that violate the social or cultural norms of a community. Sociologically, it is seen as a phenomenon that can reveal the internal dynamics of a society, including mechanisms of social control and the maintenance of norms.

Émile Durkheim, one of the founding fathers of sociology, argued that deviation is inevitable and necessary for social progress, as it challenges and modifies existing norms.[3] Robert K. Merton, with his theory of anomie, expanded Durkheim's ideas to explain how tensions between cultural goals and the socially acceptable means to achieve them can lead to deviance. Merton identifies five modes of adaptation: conformity, innovation, ritualism, retreat, and rebellion. Leaving aside conformity, which is linked to a conservative approach to theater studies, we consider innovation, for instance. It occurs when individuals accept the goals of society but use "illicit" means to achieve them, which translated to the theatrical realm can be identified with the experimentation of the technological device, from a technique-related instrument to a means of artistic communication.[4]

The search for a new techno-ritualism, on the other hand, has led to a reconsideration of the concept of community and relationship. The integration of technology in theater has profoundly transformed the concept of audience engagement, evolving from a traditional notion of "community" (commùnitas) to a more dynamic and interactive one. Henry Jenkins, in his seminal work *Convergence Culture: Where Old and New Media Collide*,[5] highlights how digital platforms enable greater participation and interaction among audiences, fostering a sense of community that transcends physical boundaries. Jenkins describes how media convergence allows audiences to actively engage with content, creating new forms of cultural production and participation.

Similarly, Giovanni Boccia Artieri's studies on new publics demonstrate how digital technologies have redefined audience experiences in the performing arts. He explains that digital tools in theater not only enhance the immediacy and intimacy of performances but also enable audiences to become co-creators, sharing their experiences and interpretations in real-time. This shift has led to the emergence of a more connected and engaged audience, where the lines between producers and consumers are increasingly blurred.[6]

By leveraging digital technologies, theater practitioners can now create immersive and interactive experiences that engage audiences on multiple levels. This evolution from a physical "community" to a digital "community" reflects a broader trend in the arts, where technology serves as a bridge between traditional and contemporary forms of engagement. The works of Jenkins and Boccia Artieri underscore the importance of understanding these shifts in audience dynamics, as they highlight the potential for new forms of cultural expression and collaboration in the digital age.

We now arrive at the artistic "movement" of interest, one that corresponds to rebellion, the rejection of both established goals and means, seeking to replace them with new values and norms. Theatrical art has moved away from traditional canons while maintaining the artistic process of defamiliarization, or *ostranénie* theorized by Viktor Shklovsky.[7] This method aims to estrange familiar objects to reveal their true essence. Applied to technological means, this process distances us from social overdetermination and the reductive perception of technical objects, without necessarily implying explicit didactics about their technicality.

> Defamiliarization does not equate to ostracism, but rather serves as a means to remove the cultural and symbolic codes that technology has accumulated, allowing for a new understanding of objects. Shklovsky, in his essay "Art as Technique," argues that art must make objects strange to renew our perception of the world. According to him, our habituation to everyday objects renders them invisible and devoid of meaning. Defamiliarization, therefore, complicates form and increases the difficulty and duration of perception, generating a sense of estrangement that forces the observer to see the object in a new light.[8]

New technologies have led to a shift due to the crisis in the perception of the spectator, weakening the subjugation to the visible that has always characterized the theatrical dimension in a conceptual and ontological way, as well as in a perceptual and phenomenological manner. On stage, non-human actors in the broadest sense are brought forth, which we have seen are not objects but "things" in a pragmatic sense that act, are animated, and are processes and relationships.

In the phenomenology of Edmund Husserl,[9] "things" are seen as objects of subjective experience. Husserl introduces the concept of "noema", which is the object of the intentional act of consciousness. The "thing" is not just an external object but is understood through the act of perception that gives it meaning. Martin Heidegger, influenced by Husserl, further explores the concept of "thing" (Zeug) and its utility in the world.[10]For him, a "thing" is an entity that has a practical use, and it is through use that things reveal their being.

Relying entirely on the solidity of these things can indeed lead to human alienation. However, we have observed that the unconscious enslavement we endure under the absolute dominance of attributing reality only to the presence of visible and tangible materiality has, in some respects, faded. Contrary to crypto-technics, which hides the technical aspect of an object to make it culturally acceptable, defamiliarization removes the cultural codes of technology, revealing its essence. This process of estrangement of machines forces us to reconsider their presence and meaning outside cultural conventions.[11] The gesture of *ostranénie* allows us to see beyond the surface of technical objects and their functions, revealing a deeper aesthetic and perceptual dimension.

Through defamiliarization, artists can transform our understanding of the technical and cultural world, challenging automatic perceptions and stimulating new critical and sensory awareness.

This process, which can be described as rebellious, has the effect of liberating perception from automatism and making the ordinary extraordinary. These adaptations and transformations demonstrate how social structures can influence deviant behavior and how the misalignment between cultural goals and legitimate means can lead to forms of deviant adaptation.[12]

Why must art, by definition an act of rebellion, find a way to survive itself and not close itself off in an entropic vision of itself? Why does theater so often represent man in a very backward manner compared to the understanding we can have of him over the past century, his contradictions, his darkness, in a word, his (in)existence as an isolable, stable, and coherent individual? These attacks are no longer the work of anti-humanists, but of writers and artists who have another idea of humanism, influenced by the evolution of human sciences, or driven by a metaphysical aspiration.[13]

It is necessary, therefore, to deviate from the post- and trans-humanist idea toward an *ab humaniste* vision, a concept introduced by French critic and theater theorist Michel Corvin.[14] The term derives from the contraction of "absurd" and "humanism", and refers to a vision of theater and art that explores the human condition in a way that fuses elements of classical humanism with the absurdities of modern life. Corvin emphasized how, in the contemporary era, many traditional humanistic narratives and ideologies have been questioned or seemed insufficient in the face of the complexities of the modern world. In this context, *ab humanisme* proposes a new way of looking at human existence, one that does not reject the importance of the individual or humanistic values but at the same time accepts and integrates irrationality and uncertainty as fundamental parts of the human experience.

The French theorist of theatrical abhumanisme refers to that paradoxical outcome, emerging in contemporary performative experimentation, where man, the indispensable object and subject of theater, completely disappears from the scene to evoke and combat an aesthetic in which "there is too much of man", too anthropocentric.[15] This approach has influenced not only theater but also other art forms, pushing artists and creators to explore themes such as alienation, identity, and inner conflict in ways that break with traditional conventions and seek to capture the fragmentary and sometimes paradoxical reality of contemporary life.[16]

Only a pragmatic approach,[17] according to which the meaning of any concept is rooted in its observable consequences and practical implications, can help orient and analyze the performances previously reported. What practical effects would a technological object, considered alien to the realm of art according to our conception, have if used to generate it? Reality is understood through the effects it has on our experiences and beliefs, and to assert that an object is inanimate or unsuitable for the performative field, it must be tested in all possible ways, even through failures. Artistic investigation must emulate the actions most appropriate to the scientific field and be seen as processes aimed at achieving a stable state of belief through continuous examination and validation of hypotheses, focusing on practical effects and empirical verification. Thus, paradoxically, a pragmatic approach to a deviant phenomenon like abhumanisme offers a method to resolve conceptual confusions and identify meaningful statements.

4.2 Space, Time, and Reality

According to the statement of the German scientist Friedrich Kittler, "Media determine our situation".[18] In other words media constitute the infrastructural basis and transcendental condition for full experience. This notion posits that media technologies shape and condition human perception, thought processes, and cultural practices by mediating and structuring the ways in which information is accessed, processed, and understood. Kittler's media theory suggests that the materiality of media technology—the physical forms and technical functionalities of media devices—profoundly influences and even dictates the nature of human interaction and cognition (Kittler, 1999). This aligns with the broader media ecological perspective that technological environments have a formative impact on societal structures and individual consciousness.[19] As the shift from oral to written cultures fundamentally transformed human memory, communication, and the organization of society. Similarly, the advent of digital media has reconfigured contemporary experiences of space, time, and identity, leading to new forms of social interaction and cultural production.[20] This understanding underscores the idea that media are not neutral conduits for information but active agents that shape the very conditions of possibility for human experience and knowledge.[21]

We are witnessing a creative and evolving process that is oriented toward relational dynamics, as the internet and virtual spaces are emerging as new anthropological environments capable of hosting the dimensions of identity and relationships. These spaces are moving beyond their initial perception as anonymous transit areas or purely instrumental zones devoid of perspective. De Kerckhove, in this regard, highlights that these spaces are no longer merely anonymous transit areas but have become new anthropological environments that profoundly influence our perception of ourselves and our social interactions. De Kerckhove's observations underscore the significant shift in how virtual spaces are perceived and utilized. They are now seen as integral to shaping human identity and fostering relationships, rather than being viewed solely as functional or transitional spaces. This transformation is indicative of the broader impact that digital environments have on social structures and individual consciousness, reinforcing their role as fundamental components of contemporary anthropological landscapes.

In the context of space-time dimensions, the idea of fragmentation and abstraction that characterized the early phase of modern technologies has been transcended.[22] For instance, quantum teleportation, which until a few years ago was considered merely a thought experiment not realizable in the real world, reached a turning point in 1997. It is now an established procedure that enables the transfer of the physical state of one particle to another, even when the particles are far apart. Quantum teleportation, the most spectacular conceivable application of the entanglement phenomenon, alongside state superposition, represents one of the most intriguing aspects of quantum mechanics. It is fascinating to compare this phenomenon to the integration of extended realities (XR) in the performative field. Consider avatars, holograms, cyberspace, and the metaverse—these are all new entities involved in the performances being analyzed.[23]

Before and after the pandemic, in a context of advanced digital ecology, numerous hybrid and hypermedia artistic initiatives have led to a significant reevaluation of online space, prompting a profound revision of the concepts of "light theatricality", simulated reality, and multimedia performance. According to Ball the complex virtual environments that have emerged as epicenters of these new creative processes and socio-anthropological experiences have marked a radical shift toward "global interconnection". This relates to the capacity of digital technologies to connect individuals and communities from different parts of the world in real-time, enabling unprecedented cultural and creative exchange.

In a continuous present, interactions and communications occur simultaneously and continuously, thanks to the constant presence and availability of digital platforms. This characteristic allows participants in cultural experiences to engage in collective activities despite differences in time zones or geographical location, contributing to a sense of global community and immediate participation. This transformation has made it evident how digital networks and interactive platforms not only modify our approach to artistic production but also redefine the modes of audience participation and perception, consolidating the role of technology as a crucial mediator in shaping contemporary cultural experiences.

The involvement of the performer or spectator is no longer tied to the material presence of the body but engages their "other being". The concept of the shaman actor[24] extends to the digital context, where performance can transcend physical boundaries through technology and action. The actor/avatar shaman becomes a figure navigating virtual realities, interacting with digital entities and virtual environments, creating a connection between the human world and the supernatural, understood as a non-human and dematerialized dimension. This parallel reinforces the anthropological idea that the actor, like the shaman, possesses the ability to transform and connect with the audience through an amplified and transfigured reality.[25]

In the current performative realm, the suspension of semantic dualities, such as true/false, real/unreal, and material/immaterial, has become more evident, leading to an exacerbated abstraction of the three Aristotelian unities. In some cases, this has resulted in an aesthetic of disappearance regarding performative corporeality. In other instances, the increasingly dominant virtualization has led to an anthropological reconsideration of the concept of the image.

Why are the eyes and all other sense organs reliable guides for human beings? Intuitively, we might say because they tell us the truth, as they act as a window onto objective reality, of which we perceive only a partial picture. However, at times, the senses are mistaken, and artists, psychologists, filmmakers, and other professionals create illusions specifically designed to deceive our perceptions. As far back as 400 B.C., Democritus stated that our perceptions related to taste or the sense of hot or cold were not reality but mere conventions. Following him, Plato compared concepts and perceptions to flickering shadows projected on the walls of a cave from an invisible reality. Since then, philosophers have debated the relationship between perception and reality. Without favoring one theory over another, we can agree that the evolution of contemporary man has led him to hide the truth: perception is no longer a window onto objective reality but an interface that conceals objective reality behind a veil of useful icons. Counter-intuitively, we might almost say that we always live within parallel realities, and the virtual world created by our senses helps us participate in the game of life.[26] Our senses are already mediated by an interface, and we see sharp details only within a small circular window, as if we were already wearing a virtual helmet beyond what technological performances ask us to wear. Thus, the experience we are led to have is not so different from the mechanism we normally enact in everyday life. We simply are not aware of adopting this perceptual mediation unless we are asked to wear or use a materially external device.[27]

So why do many spectators feel disconcerted or find this immersive performative experience unnatural? The answer includes cognitive and anthropological motivations; conscious experiences and propositional attitudes are fundamental in human nature. In one of his most influential works on this topic, *An Anthropology of Images: Picture, Medium, Body* (2011), Hans Belting delves into how images function as active entities in configuring human experiences, rather than as mere passive objects to be viewed. In this text, he explores the symbiotic relationship between images and human bodies across different epochs and cultural contexts. Images are seen as extensions of the human body, mediators between the individual and the external world, and—in the contemporary era—are fully fused and overlapped with the material corporeality of the performer.

Parallelly and consequently, in recent years, the discussion on the concept of illusion within the contemporary mediascape has intensified, particularly in relation to its relevance in digital environments such as virtual, augmented, and mixed reality technologies. These immersive environments have the ability to evoke a strong sense of integration into almost real worlds for users. Consequently, despite traditionally carrying a negative connotation, the term "illusion" is increasingly viewed in a positive light as a crucial aspect of the phenomenon of immersion. Therefore, it is considered an important goal for creators of hyper-realistic and virtual environments to pursue this effect.

No other scientific or humanistic issue fascinates us more than the nature of space and time because these two concepts form the stage on which the plot of the cosmos and humanity unfolds. Are space and time real physical entities or merely conceptual simplifications that provide coordinates for our existence? What does it mean, for example, to assert that empty space or time has a beginning, and can we manipulate them? Are space and time alterable by human intervention? Perhaps not, but technology has revealed how these two categories can be subverted and transcended.[28]

Exploring these two concepts also leads to a fundamental question: what is reality? Man has access only to his inner experiences of perception and thought, and how can he be sure that these accurately reflect the reality of the surrounding world? Perhaps the only investigative tool available is the observation of how these two categories appear clear, sharp, and objective in literature, cinema, and theater, where flashbacks and flash-forwards are clear to the audience. Equally evident in the realm of entertainment and performance is how man, through technology, can bend them to his needs.

In the academic realm, immersion in contemporary arts is configured as an anthropologically rich practice, where emotional and sensory experience is amplified through the use of advanced technological tools. These allow the exploration of innovative languages and expressive forms, delineating the concept of "anthropology of hybridization". This phenomenon is characterized by the fusion and re-elaboration of traditional cultural genres, which regenerate and reinvent themselves, giving rise to hybrid art forms still lacking a definitive nomenclature.

Metaverses are paradigmatic examples of such hybrid spaces, serving as arenas for artistic experimentation and extended creativity, challenging the boundaries of the real and the virtual in a manner that transcends previous human experiences. These augmented reality environments function as laboratories for a new virtual ethnography, where cultural interactions and identities expand beyond physical limits.

The anthropology of engagement, another crucial aspect of this new artistic reality, highlights the transformative role of the user, who becomes an active figure in artistic co-creation, going beyond mere interactivity. This process is rooted in the concept of the "participatory body", where the physical body and the cultural identity of the user converge in art, promoting participation that manifests both in the physical and virtual dimensions. The user thus assumes an anthropological role as a co-creator, actively influencing and collaborating in the creative processes in a continuous dialogue between artistic creation and human experience.

In simulated virtual environments, people experience a strong sense of presence (illusion of place) and react to what they perceive as if it were real (illusion of plausibility).[29] At the same time, they remain perfectly aware that they are not "truly" there and that events are not "actually" happening. This conflict between knowing and perceiving can be considered a new form of aesthetic illusion.[30]

The metaverse certainly provides a platform for artists to transcend the limits of physical space and materiality, offering an expanded canvas for their imagination. However, being a fictitious space dependent on the functioning of a machine, it can also generate frustration when the viewer's expectations are not met. We are thus witnessing a new way of operating and creating within a community that evokes an illusion of reality. If the metaverse offers an immersive and interactive experience that can evoke a sense of presence and engagement, the illusion of reality enacted within it can be both enticing and challenging. That said, it must be recognized that being a digital construct, there are limitations and discrepancies between the virtual experience and the physical world.

Let us now connect to an apparently distant reflection by Roger Caillois, who opens his renowned essay on animal mimicry with a focus on the concept of "distinction", recognized as one of the strongest and most innate drives in humans, expressing itself as a safeguard of one's being distinct from the environment, like a primordial organism. The boundary that delineates our being has always been the skin that contains organs and creates the concept of a unique and singular body. This assumption, which seems to distinguish us, falters when our body can expand and dematerialize in favor of creating a second body, experienced as an extension of the first organic one. We thus understand how this tension toward an escape from this biological container leads to experiments similar to those presented in the previous chapters.

Let us now connect to an apparently distant reflection by Roger Caillois, who opens his renowned essay on animal mimicry with a focus on the concept of "distinction", recognized as one of the strongest and most innate drives in humans,[31] expressing itself as a safeguard of one's being distinct from the environment, like a primordial organism. The boundary that delineates our being has always been the skin that contains organs and creates the concept of a unique and singular body. This assumption, which seems to distinguish us, falters when our body can expand and dematerialize in favor of creating a second body, experienced as an extension of the first organic one. We thus understand how this tension toward an escape from this biological container leads to experiments similar to those presented in the previous chapters.

4.3 Traditional and New Trajectories in Theatrical Anthropology

It has been a comparison between shifting fields that continuously expand and move to reflect the changes in societies and cultures. Within the theater-anthropological interplay, privileged tools and pathways have been created to rethink the ways a culture operates and acts in terms of relational representation.[32] Indeed, for the human and social sciences, theater has always been a reference point due to its ability to generate metaphors suitable for the contemporaneity in which it lives. Theater is a recognized mode through which society reinterprets itself, to see and recognize itself.[33]

> Cultural performances are the ways in which the cultural content of a tradition is organized and transmitted on particular occasions through specific media. Thus, these performances are specific and particular manifestations (examples) of culture, outside of which culture becomes an abstract counter-war. Moreover, if culture is the giving of the performance, then culture is what is given to an audience or to the theoretical external observer who joins it.[34]

To this observation, Milton Singer adds that the techniques and results attributable to the performing arts also impose themselves as constitutive of cultural performances. This recognition implied for the anthropologist the necessity to bring into play and also question the articulations of the parameters of theatricality and spectacle that he knew and used, which could be crucial for understanding how specific cultural content and values are transmitted and communicated, as well as social processes and cultural changes.[35] Although this reasoning may seem somewhat convoluted and, in some respects, redundant, it helps to understand how the reciprocity established between theater and anthropology has led to the idea that performative media considered in themselves can correspond to various forms of cultural performance. They can offer the anthropologist the opportunity to access an emerging meaning in the cultural and social event of which he is a spectator and observer.[36] Every choreography, theatrical or musical performance can thus become significant for the anthropologist as well as for the theater scholar, provided that these gestures can be inscribed in their realization and production as carriers of culture as well as carriers of the performative aesthetic status of the event.

If typically the phenomena of singing, storytelling, impersonation, and dance in traditional anthropological practice were associated with the description of ritual and its mythical, religious, or sociological meanings, relating to Singer's idea and what has been noted so far, we can confirm that they have acquired the legitimacy of "acting culturally". In this perspective, not only the forms of human language become important and decisive but also the performative metalanguage on which the reciprocal spectatorship that develops within the "cultural" performance depends.[37]

Theater and cultural anthropology, taken separately, have nothing in common either in terms of object or method of thought. However, there have been and are some individuals who migrate, sharing both affiliations, and their physical or intellectual migration produces knowledge.[38]

Why study performances from a theatrical perspective where the human is not simply deviated but dematerialized,[39] delegated, or even replaced by a mechanical and synthetic agent? Because in this performative experience, a person sees the event, sees themselves observing the event, and sees (sometimes themselves watching others who are observing the event) and who perhaps see themselves observing the event.[40] Thus, we have the performance, the performers, the spectators, the spectator of spectators, and the self that watches itself, which can be a performer or spectator, or a spectator of spectators. "It is this layering of seeing" that persists, "that radically distinguishes animal play, animal art, animal ritual, animal symbolism, animal communication, animal thought, from their human counterparts".[41]

The gap noted in the 1990s between theater and theater studies on the one hand and the development of cultural performance studies on the other was largely attributed to the disciplinary defense policies of theater studies,[42] as if they had to fight to assert their autonomy and insist on a certain separation from other studies. In reality, a sensible interdependence of a connected series of disciplines has gradually been sought, as well as the role of performance in the production of culture in its broadest sense.[43] It was later seen how the often-mentioned Geertz attributed significant recognition to theater and theater studies that went beyond their subordination to the social sciences.[44]

The convergence between theater and anthropology has been part of what Geertz calls the "collapse of classifications"[45] a process in which different disciplines clash with each other in an effort to overcome traditional dichotomies between theory and practice, mind and body, and in this context of analysis, between organic and mechanical, human and artificial. Theater has questioned itself starting from the problems addressed by the social and hard sciences, especially in more contemporary times, and has often sought answers capable of translating the theory of those sciences into its practice. This phenomenon is referred to by the selected and analyzed performances, in which the conceptual necessity of differentiating between theater and ritual manifests through the use of unprecedented and hybrid creative practices. The presence and concept of corporeality, for example, central to a more traditional theatrical anthropology like that founded by Eugenio Barba,[46] are relocated in the new social and technological dimension of theater, only to be dispersed again in everyday "agency". The body, no longer solely human but in our case digital or robotic, is not only a guarantee of emerging meaning but also a bearer of culture and, therefore, even in its paradoxical absence as a human or material entity, deposits a cultural concept in its performative action. To understand the chosen performative phenomena used as examples, it is necessary to choose schematic structures that emerge from concrete and corporeal experience rather than enunciative ones.

We have seen how theatrical anthropology is the most suitable means of studying these phenomena of contamination between human and non-human, precisely because traditional or conservative theater studies have so far looked with distrust and reluctance at what is no longer a mere transient experiment but an aesthetic of the artificial that has developed and consolidated significantly over the last twenty years and reached its peak during the COVID-19 lockdown period. Thus, the statements "if there is a robot, it is not theater" or "there must be physical co-presence in the performative experience" fall apart. This aims to show that theater can be a place where art and science can meet and dialogue, and it is a privileged place for experimentation.

I believe that the performing arts, dance, or theater can be an appropriate place for experimenting with this human/machine interaction. Certainly, the dimension of theatricality to which I refer is connected to Richard Schechner's concept of performance.[47] He has sought to demonstrate that if theater is the matrix context for many of the questions that the human and social sciences pose to individual relational action, those same questions constitute contexts for the inquiries theater poses to itself. To achieve this, he retraced the socio-anthropological reasons for the relationship between theater and society and considered cultural performances from a predominantly theatrical perspective. The notion of performance he used has returned to being readable based on a fabric of theatrical reasons that neither deny nor fear socio-anthropological ones. In his study of performance theory, he articulated the notion of performance as both a theoretical and descriptive construct positioned between theater and society/culture and theater and social sciences.[48]

The concept of performance has served theater

> to limit the danger of becoming a document of itself or of being crushed towards its own experiential creative core. For anthropology, it has served to modify the notions of artifact, text, document, and to work around the link between experience, expression, and meaning[49]

The advantage offered by the notion of performance, as evidenced by Schechner's work, lies in allowing an articulated consideration of the difference between scenic action and social action without denying their continuity. What aligns most with what has been analyzed so far is that within the concept of performance viewed from this perspective, there is something designed to be represented, but not necessarily central or an endpoint, whereas the arbitrary mobile boundaries within which it occurs are. Thus, performance cannot function without a direct reference to the society of which it is a part, and it seems entirely natural that contemporary performing arts have integrated emerging technologies into their processes and languages.

The performing arts are at the forefront of this epochal transition toward a "New Humanism", characterized by unprecedented aesthetic paths, interactions between organic and synthetic beings, and posthuman creative processes. Through the various analyses developed in this volume, an attempt to synthesize an intertwined journey in which technological research and artistic imagination, art and science, pursue uncharted paths to surpass the limits of the human, we arrive at an open-ended conclusion. By reversing perspectives, one can think of the technological world—and objects—as functioning and interpenetrating entities, akin to a world of human bodies and subjects, even in terms of intentionality and significance. Instead of mastering technological devices, the postmodern human has begun to design them in their own image, making them an extension of their body and mind. This is one of the many possible semiotic conversions, from thing to object, from object to fully fledged semiotic actor. We are now experiencing a phase of developing an awareness, distinct from consciousness, of the machine, in which the performing arts prove to be a fertile field for experimentation and implementation, as well as an anthropological observation of this rapid and ongoing evolution.

We could, maybe, conclude by saying that we are so accustomed to making the distinction between the possible and the real, between essence and existence, that we fail to realize that these notions, which seem so obvious, are neither opposed nor intimately intertwined. In fact, rather than focusing on what technology can do, it may be time to focus on ourselves, on our environment in which technology fits, on how we interact with technology, as in sometimes dystopian scenarios.

In other words, it is by articulating presence and absence that the 4.0 actor virtualizes and guarantees the space of lack necessary for the renewal of infinite forms of presence. This occurs through a process of subtraction, with what was not there before and is unknown; the lack of a relationship with what is traditionally defined as theatrical art discourages the eye but not the spirit of the spectator, who perceives the "emanation" of a new presence and the birth of a performative language in its formation, in its continuous becoming other. The stage, as we have seen, is no longer simply the space that surrounds the subject but becomes a place that 'feels,' a living environment traversed by intensities that emerge according to different gradations to the perception of those who participate in it. Technologies reveal themselves as both medium and message at the same time, no longer just tools capable of acting on traditional theatrical signs but as subjects.

Radical, rapid, and pervasive changes, such as those represented by digital transformation, inevitably cause imbalances, present challenges in outlining development strategies, and bring with them negative externalities (e.g., this is not theater, if there is no human element it cannot be called art, machine creativity will replace human creativity, etc.). To seize the opportunities of change, it is necessary to govern the transformation. Most deterministic approaches to digital and technological matters tend to view technology as an element that has effects in itself—generally interpreted in terms of disruption and radical transformation—that derive from technical possibilities without highlighting (and sometimes without even realizing) that technological innovation, the forms it takes, and the effects it has constitute a process strongly influenced by human choices and decisions, marked by intentionality and dynamics of power.[50]

> Performances are not simple reflectors or expressions of culture or even of cultural change, but may themselves be active agents of change, representing the eye by which culture sees itself and the drawing board on

> which creative actors sketch out what they believe to be more appropriate or interesting life projects.[51]

Every form of art can indeed be seen as an epistemological metaphor, as Umberto Eco claimed, if not as a substitute for scientific knowledge. This means that in every era, the way in which art forms—in our case, performative arts—are structured reflects through various processes (similarity, metaphorization, negativization, etc.) the way in which science, as well as culture, view reality.[52]

Let us now return to the introductory question of this volume, namely, what is theater, to which we will add "of today". It is undoubtedly—borrowing the words of Ortega y Gasset—innumerable different things that are born and die, vary, and transform to the point of not resembling each other at all.[53] To find a suitable answer for the historical moment, albeit not definitive, we should adopt a reversed perspective of analysis. In anthropology, the reversed perspective, which challenges conventional and ethnocentric notions, proves to be an essential tool for decoding the phenomenon of technologization in theater. This approach, highlighted in the studies of Claude Lévi-Strauss, underscores how different cultures interpret reality in ways that may seem alien or inverted compared to the dominant Western viewpoint, yet remain coherent and rational within their cultural contexts.[54] In the context of theater, this perspective allows for an examination of how technology functions not merely as a set of tools but as a dynamic actor that co-creates meanings and influences performances.[55]

By adopting a reversed perspective, one can observe that the integration of technology in theater is not merely an aesthetic addition but a fundamental transformation of performative practices. Technological objects, through this lens, are seen not only as functional instruments but as performative entities that actively participate in the construction of the theatrical reality. This approach deconstructs the traditional dichotomy between the human and the technological, revealing how both interact in a complex and interdependent manner.[56]

The reversed perspective also enhances the understanding of how digital technologies influence audience perception and interaction with performances. Instead of perceiving technology as an external or supportive element, this view acknowledges it as an integral part of the theatrical experience, capable of altering traditional dynamics of presence, space, and time. This approach not only enriches anthropological analysis but also offers new interpretative frameworks for contemporary theatrical criticism and practice, fostering a deeper comprehension of the interactions between culture and technology.[57]

4.3.1 To Avoid Concluding…

In the past two decades, the union between art and technology, while not a new phenomenon in the live performance scene, has intensified significantly.

This relationship has resulted in artistic contaminations ranging from the integration of virtual reality (VR) to mobile phones, from the use of sensors to video games, and from artificial intelligence (AI) to robotics. Thus, a process of technological 'theatrical decentralization' has gradually emerged, attracting new audience targets and contributing to the formation of new audiences and connected communities. The hybrid and transmedia nature acquired by the performing arts has helped to enhance the existing cultural and territorial heritage and has triggered, through new forms of "glocal" access and fruition of the performing arts, a sort of technological democratization of art.

The relationship between body and technologies can be seen as a field of tensions and possibilities, where technologies can both amplify and limit performative capabilities and the perception of the body. This invitation to explore such dynamics with an open yet critical mind aims to enhance the creative interaction between body and technology without losing sight of social and political implications. Could it be legitimate to think that these transformations, elaborated on a theoretical level and outlined in my brief analysis today, are linked to the change of era, to the conceptual and philosophical categories of reference that are changing? Such as those of reality, possibility, chance, and necessity?

It is not a matter of discussing and making epochal changes according to a rhetoric of continuous fracture mutation typical of the traditional and historicist approach of theater studies, and it is not necessary to resort to the post-human to become less anthropocentric. In a subversive manner, Artaud, like other re-theatricalizers of 20th-century theater, had already affirmed that traditional theater fails to capture the essence of human experience and the subconscious. He proposed a form of theater capable of breaking down the barriers between actor and audience, between reality and illusion.

Thanks to new technologies, the temporal idea is surpassed by that of an absolute present, focusing on the instant and emphasizing the episodic nature of time. This concept is also not new but is linked to the ephemeral value of the *hic et nunc* of theater. However, today, the temporal perception linked to multimedia fruition is brought to an expansion of which we are often unaware but are victims. Art is a gateway to the hermeneutic nature (in a broad sense) of humanity, and the aforementioned Fabrizio Cruciani stated that "thinking about space" means imagining it as a mental content variable between subjective and objective perceptions, constituting a "cultural convention that becomes an active element of artistic expression, both in its vision-building and in its determination as an environment, a place of expressive possibilities".[58] In our case, sharing Cruciani's idea, space is completely illusory and allusive, and the void takes shape and meaning and seeks a definition that identifies it before being filled. We analyzed a type of theatrical performance in which the stage and the objects populating it refer more to cultural conventions than to the specificity of the event, sometimes acquiring their own autonomy.[59]

The key to interpreting these new types of performances is linked to the ambivalence of the transformative power of new technologies, functioning as a metaphorical *God Janus*,[60] the Roman deity with two faces looking in opposite directions, symbolizing passages, transitions, and duality. The digital imagination evoked by them allows for an expansion of expressive and communicative possibilities. However, this transformative power also brings significant risks. As Sherry Turkle highlighted, digital technologies, while facilitating connection and creativity, can also create new forms of alienation and detachment.

Rather than estrangement, it is more appropriate to recognize a dimension of otherness on the new theatrical stage, going beyond recognition that occurs only between determined identities and which instead relies on intra-action. According to Haraway, we humans are "compost" cum-positi, in continuous formation and hybridization.[61] To perceive ourselves as such, we must go against the natural tendency of promethean ego/subject human representation.[62] Instead, a different humanity and a different conception of art, in our performative interest, are needed. Amidst the chaos of a loss of perceptual control, we must work on recognizing perception as a manifestation of a consciousness linked to experience and thus to awareness. This means detaching from perception and conception that if I perceive, represent, and grasp everything, I dominate through my perceiving.

From an anthropological perspective, this shift necessitates a reconceptualization of human identity and social interactions. Embracing the hybrid and fluid nature of our existence fosters a more inclusive and interconnected understanding of humanity. This perspective is consistent with the principles of performative anthropology, which views human actions and cultural expressions as dynamic and evolving processes. In the theatrical context, such an approach encourages artists and audiences to explore the boundaries of identity and otherness, cultivating a space where diverse experiences and perspectives can coexist and interact.[63]

Theatrically, the integration of this philosophy transforms the stage into a site of continuous negotiation and redefinition of the self. It challenges traditional narratives and encourages experimental forms that reflect the complex, intertwined realities of contemporary life. By prioritizing intra-action over mere interaction, theatre practitioners can create performances that resonate more deeply with the audience's sense of self and other, ultimately promoting a more profound engagement with the world,

Thus, recognizing the performative and ever-changing nature of human existence requires both anthropological insight and theatrical innovation. Such an approach enhances our appreciation of the multiplicity of identities and experiences that constitute our shared humanity, enabling the arts to continually explore and express the depths of this understanding.[64] To perceive ourselves as such, we must go against the natural tendency of promethean ego/subject human representation. Instead, a different humanity and a different conception of art, in our performative interest, are needed. Amidst the chaos of a loss of perceptual control, we must work on recognizing perception as a manifestation of a consciousness linked to experience and thus to awareness.

Notes

1 E. Fink, *Mask and Buskin*, Indiana University Press, Bloomington, 1972.
2 M. Richir, *Phantasia, Imagination, Affectivité: Phénoménologies et Anthropologies Phénologiques*, Jérôme Millon, Grenoble, 2004, p. 502.
3 E. Durkeim, *The Rules of Sociological Method*, The Free Press, New York, 1895 (original publication in French), English translation in 1982 and E. Durkeim, *The Division of Labour in Society*, The Free Press, New York, 1893 (original publication in French), English translation in 1984.
4 For an analysis of the technological medium as a tool, see G.C. Izenour, *Theater Technology*, Yale University Press, New Haven, CT, 1997. For an in-depth exploration of how theater can be used to communicate science and technology, transforming technical tools into artistic means of expression, see E. Weitkamp, C. Almeida, *Science & Theatre: Communicating Science and Technology with Performing Arts*, Emerald Publishing Limited, Leeds, 2022.
5 See H. Jenkins, *Convergence Culture: Where Old and New Media Collide*, New York University Press, New York, 2006.
6 As detailed in his book G. Boccia Artieri, *Gli effetti sociali del web: Forme della comunicazione e metodologie della ricerca online*, FrancoAngeli, Milano, 2015.
7 V. Chklovski, *L'art comme procédé*, in T. Todorov, *Théorie de la littérature*, Le Seuil, Paris, 1965, pp. 76–97.
8 T. Arnulf, *Emerging Technological Practices in Theater and Installation*, in Appareil [Online], 21 | 2019, published online on 15 July 2019, http://journals.openedition.org/appareil/3093, [accessed 2 June 2024], http://journals.openedition.org/appareil/3093; doi: https://doi.org/10.4000/appareil.309, https://doi.org/10.4000/appareil.3093.
9 E. Husserl (1900–1901), *Logical Investigations*, translated by J.N. Findlay, Routledge, London and New York, 2001. See also W. James (1907), *Pragmatism: A New Name for Some Old Ways of Thinking*.
10 See M. Heidegger (1927) *Being and Time*, translated by John Macquarrie and Edward Robinson, Harper & Row, New York, 1962.
11 G. Simondon, *Du mode d'existence des objets techniques*, Éditions Aubier, Paris, 2012.
12 Understanding the artistic phenomenon also involves examining how, from an anthropological perspective, deviation is analyzed in relation to specific cultural contexts, recognizing that norms vary significantly among different cultures. Mary Douglas, for instance, studied how societies manage the concept of "impurity" and how deviant behaviors threaten the boundaries between pure and impure. Victor Turner analyzed rites of passage, demonstrating how societies treat deviation as a transitional phase in the life cycle. For an in-depth exploration of how theater can be used to communicate science and technology, transforming technical tools into artistic means of expression, see: M. Douglas, *Purity and Danger: An Analysis of Concepts of Pollution and Taboo*, Routledge, London, 1966; V. Turner, *The Ritual Process: Structure and Anti-Structure*, Aldine Publishing, Chicago, 1969.
13 M. Corvin, *L'homme en trop. L'abhumanisme dans le théatre contemporain*, Les solitaires Intempestifs, Besançon, 2014, p. 12.
14 M. Corvin, *L'homme en trop. L'abhumanisme dans le théatre contemporain*, Les solitaires Intempestifs, Besançon, 2014, p. 12.

15 E. Fuoco, *Né qui, né ora: peripezie mediali della performance contemporanea*, Ledizioni, Milano, 2022, p. 165.
16 An important reference is undoubtedly the thought of Antonin Artaud, particularly in his works *Pour un théâtre avorté*, Gallimard, Paris, 2004, and *Le Théâtre et son double*, Gallimard, Paris.
17 See M. Girel, Pragmatic clarifications and dispositions in Peirce's How to Make our Ideas Clear. *Cognitio: Revista de Filosofia*, PUCSP, Sao Paulo, 2017, 1 (18), pp. 45–67.
18 F. Kittler, *Gramophone, Film, Typewriter*, Stanford University Press, Stanford, CA, 1999.
19 See M. McLuhan, *Understanding Media: The Extensions of Man*, McGraw-Hill, New York, 1964.
20 See G. Winthrop-Young, *Kittler and the Media*, Polity, Cambridge, 2011.
21 See P.J. Durham, *The Marvelous Clouds: Toward a Philosophy of Elemental Media*, University of Chicago Press, Chicago, IL, 2015.
22 See M. McLuhan, *Understanding Media: The Extensions of Man*, McGraw-Hill, New York, 1964.
23 See A.C. Dalmasso, *Le corps, c'est l'écran. La philosophie du visuel de Merleau-Ponty*, Mimesis, Milano, 2018.
24 See G. Ottaviani, *L'attore e lo sciamano*, Bulzoni, Roma, 1984.
25 See G. Ottaviani, *L'attore e lo sciamano*, Bulzoni, Roma, 1984.
26 D. Hoffmann, *The case against Reality. How Evolution Hid the Truth From our Eyes*, Penguin, 2019.
27 D. Hoffmann, *The case against Reality. How Evolution Hid the Truth From our Eyes*, Penguin, 2019.
28 See B. Green, *The Fabric of Cosmos: Space, Time and the Texture of Reality*, Penguin Editor, London, 2004.
29 See M. Slater, Philosophical transactions of the Royal Society of London, in *Series B, Biological sciences*, 364(1535), 3549–3557, 2019, https://doi.org/10.1098/rstb.2009.0138.
30 See T. Koblížek (ed.), *The Aesthetic Illusion in Literature and the Arts*, Bloomsbury, London, 2017.
31 J.P. Robinson, *The Myth Of Man: Hidden History and the Ancient Origins of Humankind*, CreateSpace Independent Publishing Platform, 2018.
32 G. Ottaviani, *Lo spettatore, l'antropologo, il performer*, Bulzoni, Roma, 1999, p. 26.
33 E. Goffman, *Frame Analysis: An Essay on the Organization of Experience*, Harvard University Press, Cambridge, MA, 1974, pp. 87–88.
34 This line of analysis of cultural performance has been developed by Giulia Ottaviani in her volume the spectator, the anthropologist, the performer. In particular, for Milton's theory, see C.Bell, *Ritual Theory, Ritual practice*, Oxford University Press, New York, 1972, pp. 70–71.
35 G. Rosen, *Economic Development and Cultural Change*, 22(1) , 140–44, 1973. http://www.jstor.org/stable/1152891, pp. 76–77.
36 G. Ottaviani, *Lo spettatore, l'antropologo, il performer*, Bulzoni, Roma, 1999. pp. 34–36.
37 Ivi, p. 36.
38 F. Taviani, *Pensare lo spettacolo,* in R. Cuppone (ed.), Pensare il teatro, Titivillus, Pisa, 2024, pp. 83–114.
39 J.F. Lyotard, L'opera come propria prammatica, in E. Mucci, P.L. Tazzi (eds), *Teorie e pratiche della critica d'arte atti del convegno*, Feltrinelli, Milano, 1979, pp. 88–109.
40 R. Schechner, Performance Studies: The Broad Spectrum Approach, in *The Drama Review*, 3(XXXII), 4–6, 1988.

41 R. Schechner, Performers and Spectators Transported and Transformed, in *The Keyon Review*, 4(III), 1981.
42 J. Dolan, Geographies of Learning: Theatre Studies, Performance and 'Performative', in *Theatre Journal*, 45, 417–441, 1993.
43 J.G. Rennelt, J.R. Roach, *Critical Theory and Performance*, The University of Michigan Press, Ann Arbor, 1992, pp. 5–6.
44 C. Geertz, *The Interpretation of Cultures*, Basic Books, New York, 2017.
45 See L. Cimmino, A. Santambrogio, *Antropologia e interpretazione. Il contributo di Clifford Geertz alle scienze sociali*, Morlacchi editore, Perugia, 2004, pp. 34, 35.
46 E. Barba, N. Savarese, *A Dictionary of Theatre Anthropology: The Secret Art of the Performer*, Taylor & Francis Ltd, 2005.
47 See R. Schechner, Performance Studies: The Broad Spectrum Approach, in *The Drama Review*, 3(XXXII), 4–6, 1988.
48 G. Ottaviani, *Lo spettatore, l'antropologo, il performer*, Bulzoni, Roma, 1984, p. 96.
49 G. Ottaviani, *Lo spettatore, l'antropologo, il performer*, Bulzoni, Roma, 1984, p. 97.
50 R.C. Gill, A. Pratt, In the Social Factory? Immaterial Labour, Precariousness and Cultural Work, in *Theory, Culture and Society*, 25(1), 1–30, 2008.
51 V. Turner, *Anthropology of Performance*, Paj Publication, New York, 1986, p. 76.
52 U. Eco, *Opera aperta. Forma indeterminazione nelle poetiche contemporanee*, Bompiani, Milano, 1962, pp. 50–51.
53 J. Ortega, Y. Gasset, *Idea del teatro. Un accenno* [1946] in Id. Idea del teatro, Medusa, Milano, 2006.
54 C. Lévi-Strauss, *Tristes Tropiques*, Plon, Paris, 1955.
55 B. Latour, *Reassembling the Social: An Introduction to Actor-Network-Theory*, Oxford University Press, Oxford, UK, 2005; D. Haraway, D.J. Simians, *Cyborgs, and Women: The Reinvention of Nature,* Routledge, 1991.
56 T. Ingold, *The Perception of the Environment: Essays on Livelihood, Dwelling and Skill.* Routledge, 2000.
57 S. Turkle, *Life on the Screen: Identity in the Age of the Internet*, Simon & Schuster, New York, 1995.
58 F. Cruciani, *Pensare lo spettacolo,* in Ferdinando Taviani, *Pensare lo spettacolo*, in Roberto Cuppone (ed.), Pensare il teatro, Titivillus, Pisa, 2024, p. 15. R. Cuppone, Pensare il teatro, cit. p. 15.
59 Ibidem.
60 The concept of the "God Janus" helps to visualize this intrinsic duality of new technologies in performance. Janus represents new beginnings and transitions, but also duality and ambiguity. Digital technologies, like Janus, look in two directions: one that explores and enhances new frontiers of artistic expression and one that looks at the potential negative effects on the social fabric and fundamental human experiences.
61 D. Haraway, *Staying with the Trouble: Making Kin in the Chthulucene*, Duke University Press, Durham, NC, 2016.
62 N. Padullo, *Estetica senza (s)oggetti. Per una nuova ecologia del percepire*, Derive Approdi, Roma, 2022, pp. 14–17.
63 See V. Turner, *The Anthropology of Performance*, PAJ Publications, New York, 1987.
64 See D.J.Gunkel, J.J.Wales, Debate: What Is Personhood in the Age of AI?, in *AI and Society*, 36, 473–486, 2021.

Index

Note: *Italic* page numbers refer to figures and page numbers followed by "n" denote endnotes.

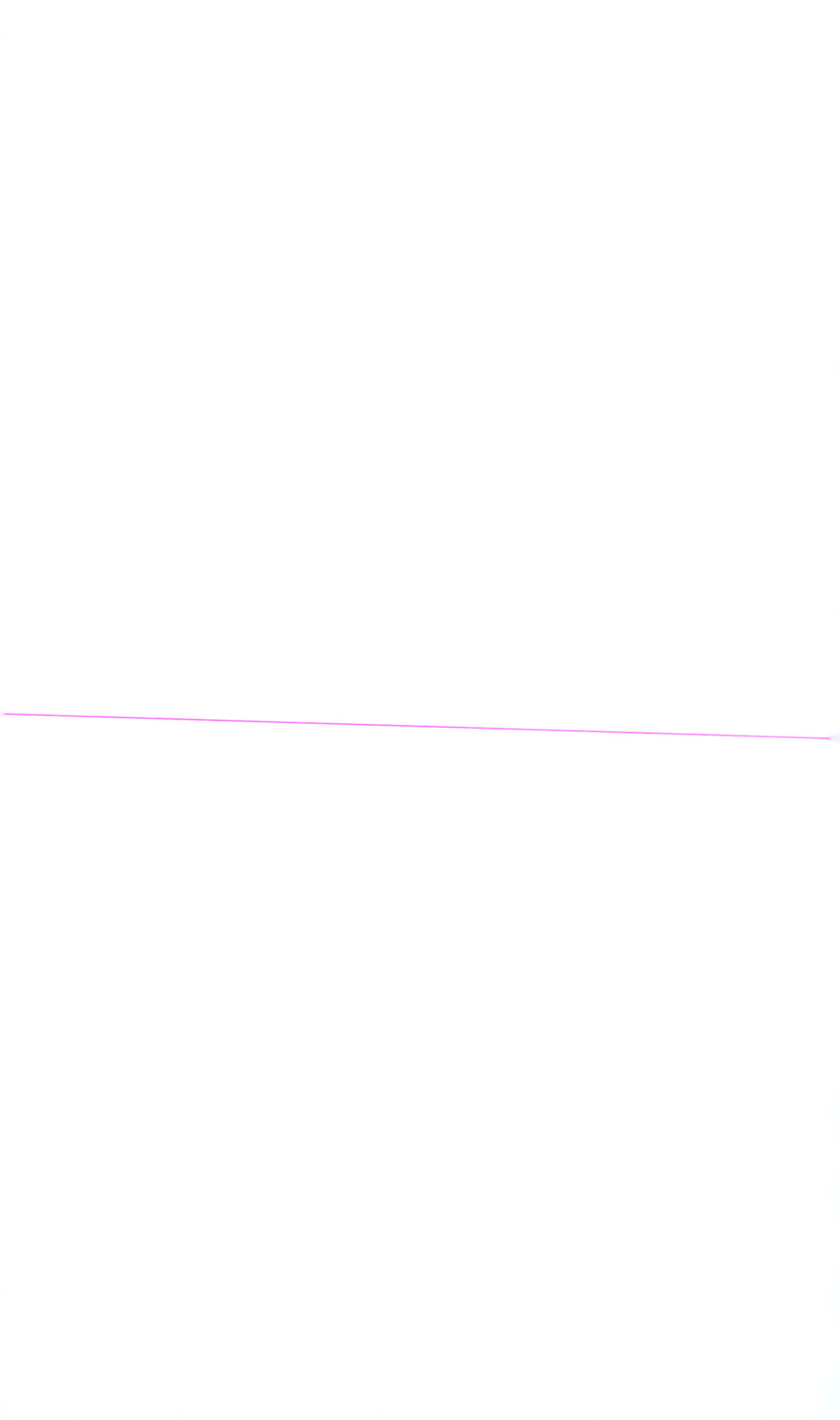

For Product Safety Concerns and Information please contact our EU representative GPSR@taylorandfrancis.com
Taylor & Francis Verlag GmbH, Kaufingerstraße 24, 80331 München, Germany

www.ingramcontent.com/pod-product-compliance
Lightning Source LLC
LaVergne TN
LVHW021603060925
820435LV00003B/27

* 9 7 8 1 0 3 2 6 7 6 9 0 6 *